Lecture Notes in Mathematics

A collection of informal reports and seminars
Edited by A. Dold, Heidelberg and B. Eckmann, Zürich

106

M. Barr, P. Berthiaume, B. J. Day,
J. Duskin, S. Feferman, G. M. Kelly,
S. Mac Lane, M. Tierney,
R. F. C. Walters

Reports of the Midwest Category Seminar III

Edited by S. Mac Lane, University of Chicago

Springer-Verlag
Berlin · Heidelberg · New York 1969

Library of Congress Catalog Card Number 70-96694 · Printed in Germany. Title No. 3262

TABLE OF CONTENTS

WHAT IS THE CENTER?

by

Michael Barr

Received January 3, 1969

Category theory was invented to define "natural". Despite this, certain very natural object constructions are not functorial in any obvious way[1]. Examples of these are completions of all kinds, injective envelope constructions and the construction of the center of a group. All except the last-named have categorical interpretations; we wish to provide one for the center. In doing so, we were motivated by considerations of obstruction theory in cohomology. The solution we derive seems right for that.

In 1. we give the basic definitions. The rest of the paper is concerned with existence: finding conditions under which every object of some category $\underline{X}$ has a center. In 2. general conditions are given, and in 3. these are applied to equational categories.

1. Basic Definition.

Let X be a group. The center $Z \subset X$ is easily seen to be the largest subgroup of X such that there exists a group homomorphism

$$Z \times X \longrightarrow X$$

whose restriction to Z is the inclusion and whose restriction to

---

[1]Actually Robert Paré, a student at McGill, has shown how all these may be made "functorial" if the mapping functions are replaced by relations.

X is the identity. Clearly, this property characterizes Z. Of course, a similar definition in an abstract category cannot make sense unless the category is pointed. Otherwise, it does not make sense to speak of restricting a map on a product to its coordinates. Accordingly, we have:

Definition 1.1: Let $\underline{X}$ be a pointed category with finite products, and let $X \in \underline{X}$. A subobject $Z \subset X$ is called central in X if there is a morphism $Z \times X \longrightarrow X$ whose restriction to Z is the inclusion, and whose restriction to X is the identity. Z is called the center of X if it is central and includes every central subobject of X.

Of course, this definition leaves the question of existence of a center wide open.

2. The Main Theorem.

Definition 2.1: A category $\underline{X}$ is called a Z-category if the following conditions are satisfied:

Z.1. $\underline{X}$ is pointed.

Z.2. $\underline{X}$ has finite projective limits.

Z.3. The "coordinate axes" $X_1 \longrightarrow X_1 \times X_2 \longleftarrow X_2$ are collectively epi for any $X_1, X_2 \in \underline{X}$.

Z.4. Any morphism $f: X \longrightarrow Y$ of X factors as $X \longrightarrow Y_0 \longrightarrow Y$ where $X \longrightarrow Y_0$ is a coequalizer (necessarily of its kernel pair) and $Y_0 \longrightarrow Y$ is monic.

Z.5. If $X \in \underline{X}$ and $\{X_i\}$ is a directed family of subobjects of X, then colim X_i exists and is a subobject of X.

Z.6. For any $X' \in \underline{X}$ the functor $X' \times -$ commutes with those inductive limits assumed in Z.4. and Z.5. This means that if $f: X \longrightarrow Y$ is a morphism which factors as $X \longrightarrow Y_0 \longrightarrow Y$ as above, then $X' \times X \longrightarrow X' \times Y_0$ is still a coequalizer (and $X' \times Y_0 \longrightarrow X' \times Y$ remains a monic). Similarly, if $\{X_i\}$ is a collection of subobjects of X, then colim $(X' \times X_i) \longrightarrow X' \times$ colim X_i by the natural map is an isomorphism.

This appears to be quite a restrictive set of hypotheses. However, many algebraic categories of interest to us satisfy them. We shall discuss this in 3.

If $X_1, \ldots, X_m, Y_1, \ldots, Y_n \in \underline{X}$ and $f: X_1 \times \ldots \times X_m \longrightarrow Y_1 \times \ldots \times Y_n$ is a morphism, then f has a matrix

$$\| f \| = \left\| \begin{matrix} f_{11} & & f_{1n} \\ & & \\ & & \\ f_{m1} & & f_{mn} \end{matrix} \right\|$$

where f_{ij} is the composition

$$X_i \longrightarrow X_1 \times \ldots \times X_m \longrightarrow Y_1 \times \ldots \times Y_n \longrightarrow Y_j .$$

The correspondence $f \longmapsto \| f \|$ is not an isomorphism as it is in an additive category, but Z.3. together with the usual properties of products insures that this correspondence is injective. If

$$X \xrightarrow{f} X_1 \times \ldots \times X_n \xrightarrow{g} X'$$

have matrices $\| f_1, \ldots, f_n \|$ and $\left\| \begin{matrix} g_1 \\ \vdots \\ g_n \end{matrix} \right\|$, we will let

$g_1 f_1 + \ldots + g_n f_n$ denote gf. The "+" does not necessarily have any real significance except that it now permits composition of maps between products to be represented by ordinary matrix multiplication. The details are familiar and will be omitted. We will frequently write down a matrix to denote a morphism, understanding, of course, that not every matrix stands for a morphism. However, a matrix with at most one non-zero map in each row always represents a morphism. For example, $\| 0, \ldots, 0, f_i, 0, \ldots, 0 \|$: $X_1 \times \ldots \times X_n \longrightarrow X$ represents $X_1 \times \ldots \times X_n \xrightarrow{\text{proj.}} X_i \xrightarrow{f_i} X$.

We are now ready to give the main result of this paper.

<u>Theorem 2.2</u>. Let $\underline{X}$ be a Z-category. Then every object of $\underline{X}$ has a center.

<u>Proof</u>. Let $X \in \underline{X}$ and $\underline{Z} = \{Z_i\}$ be the class of central subobjects of X. We must show that $\underline{Z}$ contains a largest element. First, we show it is directed. If $Z_1, Z_2 \in \underline{Z}$ and $\alpha_i : Z_i \longrightarrow X$, i = 1, 2 is the inclusion, then there is map with matrix

$$\| \alpha_i, X \| : Z_i \times X \longrightarrow X .$$

(Of course we can always write down that matrix; Z_i is central if and only if that matrix represents a map.) Now $\|\alpha_1, \alpha_2\|: Z_1 \times Z_2 \longrightarrow X$ is a morphism since it can be factored, e.g., $\|\alpha_1, X\| \begin{Vmatrix} Z_1 & 0 \\ 0 & \alpha_2 \end{Vmatrix}$. Let $P \overset{d^0}{\underset{d^1}{\rightrightarrows}} Z_1 \times Z_2$ and $Z_1 \times Z_2 \xrightarrow{d} Z_1 Z_2$ be the kernel pair of $\|\alpha_1, \alpha_2\|$ and the coequalizer of d^0, d^1 respectively. Suppose that $d^i = \begin{Vmatrix} \partial_{1i} \\ \partial_{2i} \end{Vmatrix}$, $i = 0, 1$, and $d = \|\beta_1, \beta_2\|$. By Z.4. the induced map $\alpha: Z_1Z_2 \longrightarrow X$ is a subobject, and, of course, $\alpha\beta_i = \alpha_i$, $i = 1, 2$. Also $\beta_1\partial_{10} + \beta_2\partial_{20}$ and $\beta_1\partial_{11} + \beta_2\partial_{21}$ are defined and equal which implies that $\alpha_1\partial_{10} + \alpha_2\partial_{20}$ and $\alpha_1\partial_{11} + \alpha_2\partial_{21}$ are also defined and equal. Now by Z.6.

$$P \times X \overset{d^0 \times X}{\underset{d^1 \times X}{\rightrightarrows}} Z_1 \times Z_2 \times X \xrightarrow{d \times X} Z_1Z_2 \times X$$

is also a coequalizer. The map with matrix $\|\alpha_1, X\|(Z_1 \times \|\alpha_2, X\|)$ $= \|\alpha_1, X\| \begin{Vmatrix} Z_1 & 0 & 0 \\ 0 & \alpha_2 & X \end{Vmatrix} = \|\alpha_1, \alpha_2, X\|$ coequalizes $d^0 \times X$ and $d^1 \times X$. In fact $\|\alpha_1, \alpha_2, X\| \begin{Vmatrix} \partial_{10} & 0 \\ \partial_{20} & 0 \\ 0 & X \end{Vmatrix} = \|\alpha_1\partial_{10} + \alpha_2\partial_{20}, X\|$

$= \|\alpha_1\partial_{11} + \alpha_2\partial_{21}, X\| = \|\alpha_1, \alpha_2, X\| \begin{Vmatrix} \partial_{11} & 0 \\ \partial_{21} & 0 \\ 0 & X \end{Vmatrix}$. Thus there is induced a map $\|\gamma_1, \gamma_2\|: Z_1Z_2 \times X \longrightarrow X$ with

$\|\gamma_1, \gamma_2\| \cdot (d \times X) = \|\alpha_1, \alpha_2, X\| = \|\alpha\beta_1, \alpha\beta_2, X\|$

$= \|\gamma_1, \gamma_2\| \begin{Vmatrix} \beta_1 & \beta_2 & 0 \\ 0 & 0 & X \end{Vmatrix} = \|\gamma_1\beta_1, \gamma_1\beta_2, \gamma_2\|$. Then

$\gamma_2 = X$ and $\|\alpha\beta_1, \alpha\beta_2\| = \|\gamma_1\beta_1, \gamma_1\beta_2\|$ or $\alpha d = \gamma_1 d$. Since d is a coequalizer, hence epi, it follows that $\alpha = \gamma_1$ and Z_1Z_2 is central. Of course $Z_i \subset Z_1Z_2$, i = 1, 2, since the inclusion map $Z_j \longrightarrow X$ factors through it.

Now since $\underline{Z}$ is a directed family of subobjects of X, it has a colimit Z which is also a subobject α: Z $\longrightarrow$ X by Z.5. If β_i: $Z_i \longrightarrow Z$ is the map to the colimit, then also the β_i are mono and $\alpha\beta_i = \alpha_i$. By Z.6., $Z \times X = \text{colim } Z_i \times X$ and since for each i, $\|\alpha_i, X\|$: $Z_i \times X \longrightarrow X$ is a map, there is induced a map $\|\gamma, \gamma'\|$: $Z \times X \longrightarrow X$ such that for each i, $\|\gamma, \gamma'\|(\beta_i \times X) = \|\alpha_i, X\|$. This gives $\|\gamma\beta_i, \gamma' X\| = \|\alpha_i, X\|$ or $\gamma' = X$, $\gamma\beta_i = \alpha_i$ for all i. Since also $\alpha\beta_i = \alpha_i$, the uniqueness of map extensions guarantees that $\gamma = \alpha$, so $\|\alpha, X\|$ is a map. Thus Z is central and clearly contains all central subobjects.

3. Equational Categories.

By an equational category, we mean a category $\underline{X}$ equipped with an algebraic functor U: $\underline{X} \longrightarrow$ Sets (i.e., one which is tripleable as soon as it has an adjoint). This means that if F is a functor with codomain $\underline{X}$ and S = lim UF, then there is a unique (up to isomorphism) $X \in \underline{X}$ with X = lim F and UX = lim UF. Also, if $X \subset Y \times Y$ is such that $UX \subset UY \times UY$ is an equivalence relation, then $X \rightrightarrows Y$ has a coequalizer $Y \to Z$ and $UX \rightrightarrows UY \to UZ$ is a coequalizer. If n is any set (possibly infinite), an n-ary operation is a natural transformation of $U^n \to U$. U has a left adjoint F if and only if, for each set n, the class of natural transformations of $U^n \to U$ is a proper set. (And then UFn = nat. trans. (U^n, U).) A nullary operation, also called a constant, is a natural transformation of $U^0 = 1 \to U$. A natural transformation $U^n \to U^m$ is called a projection if it is of the form U^f where $f\colon m \to n$ is a function. We say that "all operations are finite" when we actually mean that any n-ary operation $U^n \to U$ factors as $U^n \to U^{n_0} \to U$ where the first map is a projection and n_0 is a finite set. For more details of the theory of equational categories see [2].

Theorem 3.1. Let $\underline{X}$ be an equational category. Then:

1. $\underline{X}$ satisfies Z.1. if and only if there is exactly one nullary operation.
2. $\underline{X}$ satisfies Z.2.
3. $\underline{X}$ satisfies Z.3. (when $\underline{X}$ is pointed) if there is a binary operation "+" satisfying $x + 0 = 0 + x$ for $x \in\in \underline{X}$ where 0 is the base point. A tripleable category $\underline{X}$ satisfies Z.3'; $X * Y \to X \times Y$ onto, if and only if there is such a "+". (Here $*$ is the coproduct.)
4. $\underline{X}$ satisfies Z.4.
5. $\underline{X}$ satisfies Z.5. if all operations are finite. If $\underline{X}$ is tripleable, the converse holds.

6. $\underline{X}$ satisfies Z.6.

<u>Proof</u>:

1. This is well-known. Permit me to observe, however, that it requires showing that if α is an n-ary operation, then $\alpha(0, \ldots, 0) = 0$. But if this were not an equation in the system, then $\alpha(0, \ldots, 0)$ would define a new nullary operation.
2. See, for example [2], p. 87.
3. It is well-known that in an equational category, there are coproducts which we denote by *. Then Z.3. is just the statement that the natural map $X_1 * X_2 \longrightarrow X_1 \times X_2$ is an epimorphism. If there is a binary operation + with $x + 0 = 0 + x = x$, then $(x_1, x_2) = (x_1, 0) + (0, x_2)$. Each of those is clearly in the image, so their sum is. Thus the natural map is onto, and Z.3' holds. Conversely, if Z.3' holds and $\underline{X}$ is tripleable, the natural map $F1 * F1 \longrightarrow F1 \times F1$ is onto and we can find an element $\xi \in UF2 = U(F1 * F1)$ whose image in $UF1 \times UF1$ is (η, η) where η is the generator of F1. UF2 = nat. trans. (U^2, U) and the natrual transformation corresponding to ξ is the desired one. The details are left to the reader (see [2]).
4. See [2], p. 88 (called the First Isomorphism theorem).
5. This seems to be known, but as I have been unable to find a reference in the literature, I will include a proof. If $\underline{X}$ is finitary and $X, Y \in \underline{X}$, a set function $f: X \longrightarrow Y$ need only commute with finite

operations to be a morphism, since commuting with projections is automatic. Now if $\{X_i\}$ is a directed family of subobjects of X and if $\{f_i\}$, $f_i: X_i \longrightarrow Y$ is a family of maps on the direct system, let $X' = \bigcup X_i$ (set union). X' is a subobject, for if ω is an n-ary operation, n finite, and $x_{i_1}, \ldots, x_{i_n} \in X'$, I can already find X_i with each of $x_{i_1}, \ldots, x_{i_n}$ and hence $\omega(x_{i_1}, \ldots, x_{i_n})$ being elements of X_i. Similarly, the $\{f_i\}$ extends to a set map $f: X' \longrightarrow Y$ and $f\omega(x_{i_1}, \ldots, x_{i_n}) = f_\alpha\omega(x_{i_1}, \ldots, x_{i_n}) = \omega(f_i(x_{i_1}), \ldots, f_i(x_{i_n})) = \omega(f(x_{i_1}), \ldots, f(x_{i_n}))$, since f extends f_i and f_i is a morphism. Conversely, if Z.5 holds and $\underline{X}$ is tripleable, we have $n = \operatorname{colim} n_0$ where n_0 ranges over the finite subsets of n. But a left adjoint F commutes with colimits, so $Fn = \operatorname{colim} Fn_0$, certainly $\{Fn_0\}$ is directed and their union is exactly the n-ary operations which are composites of projections and finitary operations. If Fn is just this union, then this union includes all the n-ary operations.

6. In an equational category a map is a coequalizer if and only if it is surjective. Let $f: X \longrightarrow Y$ be a map. The point set image Y_1 of f is also its categorical image and we have f factoring as

$X \xrightarrow{\text{onto}} Y_0 \xrightarrow{\text{1-1}} Y$. Also $X' \ltimes -$ preserves both properties of being 1-1 and onto and so $X' \ltimes f$ factors $X' \ltimes X \longrightarrow X' \ltimes Y' \longrightarrow X' \ltimes Y$ with the first being a coequalizer and the second being 1-1, and hence the image of $X' \ltimes f$. As for the second half, if $\{X_i\}$ is a directed family of subobjects, the colim X_i is just the set theoretic union (of course all operations are finite). $\{X' \ltimes X_i\}$ is still directed and $X' \ltimes -$ commutes with set union.

Thus we have proved,

Theorem 3.2. Let $\underline{X}$ be a pointed equational category with all operations finitary and in which there is a binary operation for which the base point is a 2-sided unit. Then every object of $\underline{X}$ has a center.

Let us examine this situation more closely. If Z is the center of X and if there is a map $\tau: Z \ltimes X \longrightarrow X$ with $\tau(z, 0) = z$ and $\tau(0, x) = x$ for $z \in Z$, $x \in X$, then τ must commute with all the operations. If Z.3' holds, then it must in particular commute with the distinguished binary operation, denoted by +, appearing in the statement of theorem 3.2. Thus:

$$\tau(z + z', x + x') = \tau(z, x) + \tau(z', x') .$$

If $z' = x = 0$, this says $\tau(z, x') = \tau(z, 0) + \tau(0, x') = z + x'$ so $\tau = +$. Then $(z + z') + (x + x') = (z + x) + (z' + x')$. Then if $z = x' = 0$, we get $z' + x = x + z'$. Finally, letting $z' = 0$, we have z + (x + x') = (z + x) + x'. Thus Z is a commutative associative monoid and the operation of Z on X is commutative and associative. A modification of this result to make Z into a group has long been known in universal algebra; see for example [1] pp. 799-800.

Note: It has recently come to the author's attention that S. A. Huq [commutator, nilpotency and solvability in categories, Quart. J. Math. Oxford (2), 19 (1968), 363-389] has considered closely related concepts (with arbitrary maps rather than subobjects) except that his axioms are strong enough to make central subobjects be abelian groups (not merely monoids) but lacking continuity axioms Z.5 and Z.6 he cannot prove that centers exist.

REFERENCES

1. P. Crawley and B. Jonsson, Refinements for infinite direct decompositions of algebraic systems, Pacific J. Math. 14 (1964), 797-855.

2. F.E.J. Linton, Some aspects of equational categories, Proceedings of the conference on categorical algebra, La Jolla 1965, 84-94.

THE FUNCTOR EVALUATION

by

P. Berthiaume

Received October 31, 1968

Introduction. The purpose of this article is to show the role played in category theory by the functor evaluation and the notions of functorial density and cofinality with respect to the Kan Extension Theorems and related propositions of André [I] and Lambek [6].

The present paper is an outgrowth of work begun at the University of Chicago in the summer of 1967, while I was holding an NRC (Canada)-Nato postdoctoral fellowship. It owes a lot to Saunders Mac Lane, Jean Bénabou and Bill Lawvere, who taught me many unpublished facts of Category Theory and are directly responsible for some of the results, as I will try to indicate in the text, and specially to the former for his many suggestions and corrections concerning this article (in particular for the notion of cofinality). I realize that this paper includes many essentially well known propositions and proofs, but this is done for the sake of completeness and especially unification of the theory.

1. Preliminaries

If $\underline{C}$ is a category, it will always be understood that C, C_1, C_2, C' are objects of $\underline{C}$, and c, c_1, c_2, c' arrows in C. Functors will be described as follows:

$$F: \underline{C} \longrightarrow \underline{D}: C_1 \xrightarrow{c} C_2 \rightsquigarrow F(C_1) \xrightarrow{F(c)} F(C_2),$$

or by writing

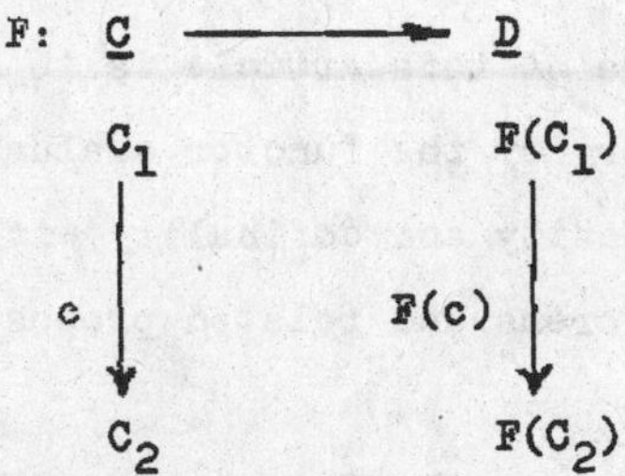

and natural transformation will be written $\eta: F \to G$. The identity endofunctor on $\underline{C}$ will be denoted by $I_{\underline{C}}$ or imply I, the identity endomorphism on C by 1_C or 1, and natural equivalences and isomorphisms of categories by $\cong$.

From now on, $\underline{A}$ will always denote a small category. If $\underline{B}$ is arbitrary, $(\underline{A}, \underline{B})$ will be the category of all functors from $\underline{A}$ to $\underline{B}$ and their natural transformations. A functor $D: \underline{A} \longrightarrow \underline{B}$ is then a diagram, and $\underline{B}$ is said to be D-cocomplete whenever the colimit of D exists in $\underline{B}$, and $\underline{A}$-cocomplete when the same holds for all diagrams $D: \underline{A} \longrightarrow \underline{B}$. We will use the well known isomorphism of categories:

(1.1) $$(\underline{B} \times \underline{A}, \underline{C}) \cong (\underline{B}, (\underline{A}, \underline{C}))$$

which still makes sense when $\underline{B}$ is not small: just write everything in words, i.e., use "metafunctors".

If F is a functor on the left in this isomorphism, $F_{\underline{B}}$ will denote be corresponding one on the right. Conversely, to G on the right will correspond $G_{\underline{B}\ltimes\underline{A}}$ on the left.

From a given category $\underline{A}$, one can manufacture a category $\hat{\underline{A}}$, the Kan subdivision category of $\underline{A}$ (cf. [5], and [12] where it is called Mor($\underline{A}$)) as follows: the set of objects of $\hat{\underline{A}}$ is the union of the set of objects of $\underline{A}$ and that of those arrows a of $\underline{A}$ which are not identity arrows, where A in $\underline{A}$ becomes $\hat{A}$ in $\hat{\underline{A}}$, and a: A $\longrightarrow$ A' in $\underline{A}$ becomes an object $\hat{a}$ in $\hat{\underline{A}}$, and the only arrows in $\hat{\underline{A}}$ are of the form $\hat{A} \xleftarrow{\grave{a}} \hat{a} \xrightarrow{\acute{a}} \hat{A}'$, $\grave{a}$ and $\acute{a}$ being new symbols (that we will omit in the future). It is obvious that $\widehat{(\underline{A}^*)}$ is isomorphic to $\hat{\underline{A}}$.

<u>THEOREM 1</u>. $\underline{C}$ is $\hat{\underline{A}}$-cocomplete iff $\underline{C}$ has pushouts and all coproducts of cardinality less or equal to that of the set $E = |\hat{\underline{A}}| - \{\hat{A} | \hat{A} \text{ in } \hat{\underline{A}}\}$.

<u>Proof</u>. Let $\underline{C}$ be $\hat{\underline{A}}$-cocomplete and $C_1 \longleftarrow C \longrightarrow C_2$ be given in $\underline{C}$. Let P: $\hat{\underline{A}} \longrightarrow \underline{C}$: $\hat{A} \longleftarrow \hat{a} \longrightarrow \hat{A}' \rightsquigarrow C_1 \longleftarrow C \longrightarrow C_2$ for all $\hat{a}$ in $\hat{A}$. The colimit of that functor is the required pushout. For any subset E' of E and set $\{C_a | C_a \text{ in } \underline{C},\ a \in E\}$, let S: $\hat{\underline{A}} \longrightarrow \underline{C}$: $\hat{A} \longleftarrow \hat{a} \longrightarrow \hat{A}' \rightsquigarrow C_a$ if $a \in E'$, and onto a fixed $C_{a'}$, $a' \in E'$, if $a \notin E'$. The colimit of S is the coproduct of the given family. Conversely the coproduct over the set E of the pushouts of $D(\hat{A}) \longleftarrow D(\hat{a}) \longrightarrow D(\hat{A}')$ for each $\hat{a}$ in $\hat{\underline{A}}$ is the colimit of D: $\hat{\underline{A}} \longrightarrow \underline{C}$. Q.E.D.

We will also make frequent use of the comma categories of Lawvere ([7]). Let us recall the definition: if $\underline{B} \xrightarrow{F} \underline{C} \xleftarrow{G} \underline{D}$, then an object in the comma category (F, G) is a triple (B, c, D), where $F(B) \xrightarrow{c} G(D)$, and an arrow $(B, c, D) \longrightarrow (B', c', D')$ in (F, G) is a pair $(B \xrightarrow{b} B', D \xrightarrow{d} D')$ such that G(d).c = c'.F(b). (Or more precisely, a quadruple (c, b, d, c')!)

Let $\underline{1}$ be the category with single object 1 and ϵ_B: $\underline{1} \longrightarrow \underline{B}$: $1 \rightsquigarrow B$. Let us denote by Ens the category of sets and 1 the set with only one element. Then one has various instances of comma categories that will be used later in the text; e.g.

(1.2) $\underline{A}/A = (I_{\underline{A}}, \epsilon_A)$, the category $\underline{A}$ over A, where $I_{\underline{A}}$ is the identity endofunctor of $\underline{A}$. $A/\underline{A} = (\epsilon_A, I_{\underline{A}})$ is then the category $\underline{A}$ under A.

(1.3) (Y, ϵ_S) where Y: $\underline{A} \longrightarrow (\underline{A}^*, \text{Ens})$ is the Yoneda functor, corresponding to $\underline{A}(\ ,\ )$: $\underline{A}^* \times \underline{A} \longrightarrow$ Ens through (1.1), and S: $\underline{A}^* \longrightarrow$ Ens. If $(\underline{A}^*, \text{Ens})_r$ denotes the full subcategory of ($\underline{A}^*$, Ens) consisting of all corepresentable functors from $\underline{A}^*$ to Ens, then (Y, ϵ_S) is the category $(\underline{A}^*, \text{Ens})_r/S$, and it is isomorphic to the category of S-copointed objects, (c.f. [10]), by the co-Yoneda lemma.

(1.4) $F\_/B = (F, \epsilon_B)$ where F: $\underline{A} \longrightarrow \underline{B}$ and B is in $\underline{B}$. Now for each B in $\underline{B}$, F generates a functor $\overline{F}(B) = \underline{B}(F(\ ), B)$: $\underline{A}^* \longrightarrow$ Ens which in turn gives $\overline{F}$: $\underline{B} \longrightarrow (\underline{A}^*, \text{Ens})$; the co-Yoneda lemma then implies that $F\_/B$ is isomorphic to $(\underline{A}^*, \text{Ens})_r/\overline{F}(B)$. Dually, $B/F\_ = (\epsilon_B, F)$.

(1.5) $(\epsilon_1, \underline{A}(\ ,\ ))$, where $\epsilon_1 : \underline{1} \to \text{Ens}: 1 \rightsquigarrow 1$. This is isomorphic to the category $\tilde{\underline{A}}$ of "twisted morphisms" of $\underline{A}$. (cf. [1]): an object is an arrow $A \xrightarrow{a} A'$ in $\underline{A}$, and an arrow $a \to a'$ in $\tilde{\underline{A}}$ is a pair of arrows (a_1, a_2) such that $a_2 \cdot a \cdot a_1 = a'$, where $A_1 \xrightarrow{a'} A_1'$.

(1.6) $(\epsilon_1, \underline{B}(\ , F(\ )))$ where $F: \underline{A} \to \underline{B}$ and thus $\underline{B}(\ , F(\ )): \underline{B}^* \times \underline{A} \to \text{Ens}$. This is isomorphic to the following category $\underline{L}_F$: an object is a triple (B, b, A) where $B \xrightarrow{b} F(A)$ and an arrow a pair (b_1, a) such that $F(a) \cdot b \cdot b_1 = b'$ where $B' \xrightarrow{b'} F(A')$. Of course (1.5) is a special case of (1.6).

Finally, we shall need the following three functors:

(1.7) $T_o: \hat{\underline{A}} \longrightarrow (\tilde{\underline{A}})^*$

$$
\begin{array}{ccccc}
\hat{A} & & A & \xrightarrow{\ 1\ } & A \\
\uparrow & & \uparrow{\scriptstyle 1_A} & \vdots\uparrow & \downarrow{\scriptstyle a} \\
\hat{a} & & A & \xrightarrow{\ a\ } & A' \\
\downarrow & & \downarrow{\scriptstyle a} & \vdots\downarrow & \uparrow{\scriptstyle 1_A'} \\
\hat{A}' & & A' & \xrightarrow{\ 1\ } & A'
\end{array}
$$

(1.8) $K_o: (\tilde{\underline{A}})^* \longrightarrow \underline{A} \times \underline{A}^*$

$$
\begin{array}{ccccc}
A & \xrightarrow{\ a\ } & A' & & (A, A') \\
\downarrow{\scriptstyle a_1} & \vdots\downarrow & \uparrow{\scriptstyle a_2} & & \downarrow{\scriptstyle (a_1, a_2)} \\
A_1 & \xrightarrow{\ a'\ } & A_1' & & (A_1, A_1')
\end{array}
$$

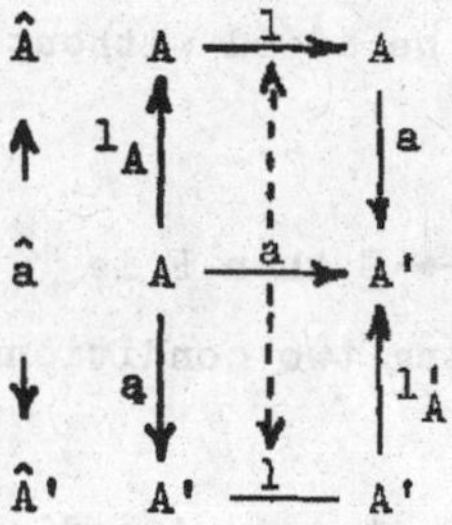

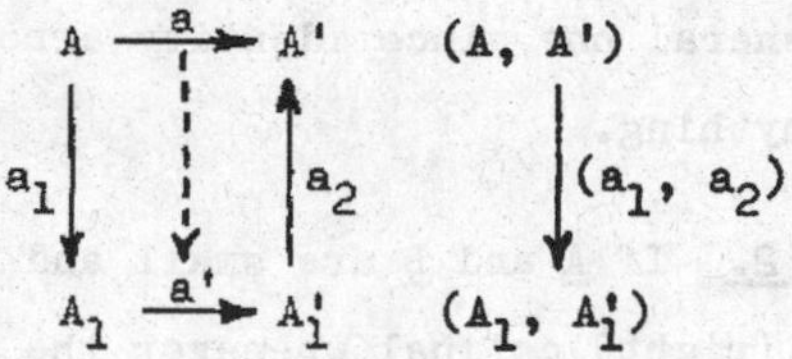

(1.9) $L_F: (\tilde{\underline{A}})^* \longrightarrow \underline{L}_F^*$

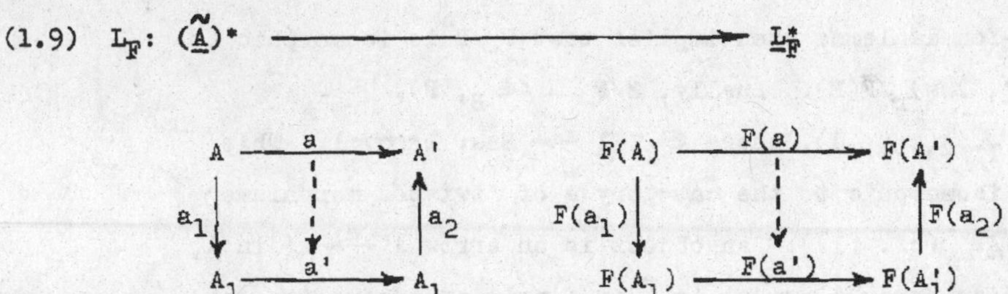

where T_o is a bijection on the objects and faithful.

2. Cofinality

This section contains a generalization of the classical notion of cofinality.

DEFINITION 1. A category $\underline{C}$ is connected whenever any two objects C and C' in $\underline{C}$ can be joined by a finite sequence of arrows in $\underline{C}$.

A typical case would be:

$C \xrightarrow{c_1} C_1 \xleftarrow{c_2} C_2 \xrightarrow{c_3} \ldots \xrightarrow{c_{2n-1}} C_{2_{n-1}} \xleftarrow{c_{2n}} C'$ and in fact this is the most general one since identity arrows can be added without changing anything.

DEFINITION 2. If $\underline{A}$ and $\underline{B}$ are small and $F: \underline{A} \longrightarrow \underline{B}$ then F is said to be (right) cofinal whenever the following two conditions hold:

(1.10) for every object B in $\underline{B}$ there is an object A in $\underline{A}$ and an arrow $B \xrightarrow{b} F(A)$ in $\underline{B}$,

(1.11) for every B in $\underline{B}$ the category B/F- is connected.

The classical case goes as follows: F is assumed to be a full embedding satisfying (1.10) and:

(1.12) for any B in $\underline{B}$ and arrows $B \xrightarrow{b} F(A)$ and $B \xrightarrow{b'} F(A')$ there exists an object A" in $\underline{A}$ and arrows $A \xrightarrow{a} A"$ and $A' \xrightarrow{a'} A"$ such that $F(a).b = F(a').b'$.

It is easy to see that the above classical version is a special case of Definition 3, while the former applied to posets $\underline{A}$ and $\underline{B}$ with F the inclusion map implies that $\underline{B}$ is directed (by 1.12) and that $\underline{A}$ is cofinal in $\underline{B}$ (by 1.10).

Let us return to the general case $F: \underline{A} \longrightarrow \underline{B}$. If $D: \underline{B} \longrightarrow \underline{C}$ then D_oF is a "sub-diagram" of D and hence there is a canonical morphism $\delta(D): \mathrm{colim}(D \circ F) \longrightarrow \mathrm{colim}(D)$ in $\underline{C}$ whenever the latter two colimits exists in $\underline{C}$. If $\underline{C}$ is $\underline{A}$ and $\underline{B}$-cocomplete this induces a natural transformation δ from the functor $(\underline{B}, \underline{C}) \xrightarrow{(F, I)} (\underline{A}, \underline{C}) \xrightarrow{\mathrm{colim}} \underline{C}$ to the functor $(\underline{B}, \underline{C}) \xrightarrow{\mathrm{colim}} \underline{C}$ where $(F, I)(D) = D \circ F$ for any diagram $D: \underline{B} \longrightarrow \underline{C}$.

<u>THEOREM 2.</u> Let $F: \underline{A} \longrightarrow \underline{B}$ be cofinal and $D: \underline{B} \longrightarrow \underline{C}$. If $\eta: D \Rightarrow C_{\underline{B}}$ is the colimit of D then $\eta_* F: D \circ F \Rightarrow C_{\underline{A}}$ is the colimit of $D \circ F$. Here $C_{\underline{B}}: \underline{B} \longrightarrow \underline{C}: B \xrightarrow{b} B' \rightsquigarrow C \xrightarrow{1} C$ and $C_{\underline{A}}: \underline{A} \longrightarrow \underline{C}: A \xrightarrow{a} A' \rightsquigarrow C \xrightarrow{1} C'$ are the constant diagrams and $(\eta_* F)(A) = \eta(F(A))$.

<u>Proof</u>. $\eta_* F$ is a natural transformation and any natural transformation $\Theta: DF \Rightarrow C'_{\underline{A}}$ can be lifted to $\overline{\Theta}: D \Rightarrow C'_{\underline{B}}$ by setting $\overline{\Theta}(B) = \Theta(A) \,.\, D(b)$ where $B \xrightarrow{b} F(A)$ exists by (1.10): $\overline{\Theta}$ is natural for if $b_o: B \longrightarrow B'$ then by (1.10) again there is a

$B' \xrightarrow{b'} F(A')$ and b and $b'.b_o$ are connected in B/F- by (1.11); the definition of $\bar{\Theta}$ is independent of the choice of the b allowed by (1.10) for if $B \xrightarrow{b''} F(A'')$ is another choice then again b and b" are connected in B/F- by (1.11) and in particular $\bar{\Theta}(F(A)) = \Theta(A)$. The naturality of $\bar{\Theta}$ implies the existence of a unique $c: C \longrightarrow C'$ such that for all B, $c.\eta(B) = \bar{\Theta}(B)$, and in particular, $c.\eta(F(A)) = \bar{\Theta}(F(A)) = \Theta(A)$. If $c': C \longrightarrow C'$ is such that for all A in $\underline{A}$, $c'.\eta(F(A)) = \Theta(A)$, then for any B, $c'.\eta(B) = c'.\eta(F(A)), D(b) = \Theta(A) . D(b) = \bar{\Theta}(B)$ and thus $c' = c$ by the uniqueness of c.

<u>COROLLARY.</u> If $F: \underline{A} \longrightarrow \underline{B}$ is cofinal, $D: \underline{B} \longrightarrow \underline{C}$ and colim D exists in $\underline{C}$, then $\delta(D)$ is an isomorphism. If $\underline{C}$ is $\underline{B}$-cocomplete then δ is a natural equivalence.

The following theorem is mentioned in [9] and attributed to Mac Lane:

<u>THEOREM 3.</u> The functor $T_o: \hat{\underline{A}} \longrightarrow (\tilde{\underline{A}})^*$ is cofinal.

<u>Proof</u>. As mentioned before T_o is onto on objects and thus the b of (1.10) is taken to be the identity, i.e., one has $a \xrightarrow{(1,\ 1)} T_o(\hat{a}) = a$ for all $A \xrightarrow{a} A'$ in $(\tilde{A})^*$. If $A_1 \xrightarrow{a_1'} A_1'$ is another object of $(\tilde{A})^*$ and $a \xrightarrow{(a_1,\ a_2)} a'$ is a morphism in $(\tilde{A})^*$, i.e., $A \xrightarrow{a_1} A_1$ and $A_1' \xrightarrow{a_2} A'$ and $a_2.a'.a_1 = a$ then the following diagram shows that (1, 1) above and (a_1, a_2) are connected in a/T_o-:

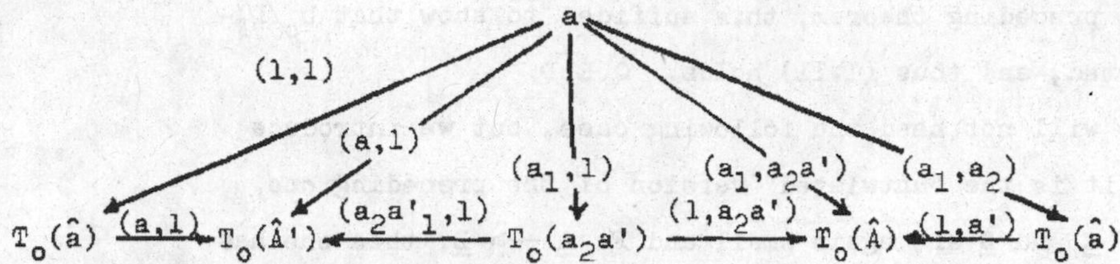

This shows that a/T_o- is connected because any other object of a/T_o- could be connected to (1,1) as above and then ultimately to (a_1, a_2).

<u>THEOREM 4.</u> The functor L_F: $(\tilde{\underline{A}})^* \longrightarrow \underline{L}^*_F$ is cofinal.

<u>Proof</u>: Recall that F: $\underline{A} \longrightarrow \underline{B}$ where $\underline{A}$ and $\underline{B}$ are small (c.f. (1.6)). If b_o: $B \longrightarrow F(A)$ is a typical element of $\underline{L}^*_F$ then one has $(b_o, 1_A)$: $b_o \longrightarrow L_F(1_A)$ (in fact L_F here plays the role of the F of Definition 2) and thus (1.10) is satisfied. Let now a_1: $A_1 \longrightarrow A'_1$, a: $A_1 \longrightarrow A$, and (b_1, a): $b_o \longrightarrow L_F(a_1) = F(a_1)$ be a morphism in $\underline{L}^*_F$, i.e., $F(a).F(a_1).b_1 = b_o$. The following diagram shows that $(b_o, 1_a)$ and (b_1, a) are connected in b_o/L_F- :

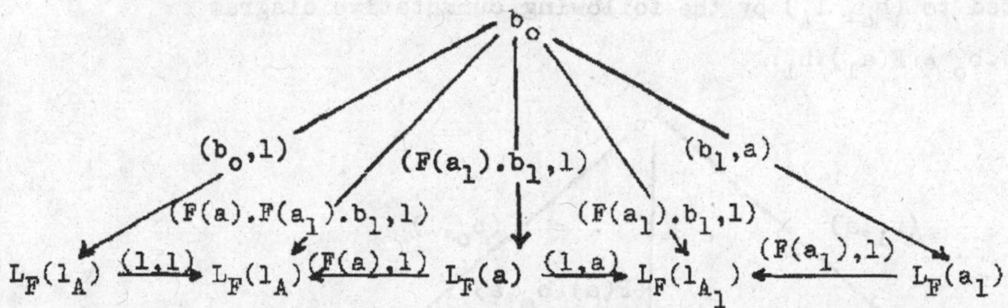

As in the preceding theorem, this suffices to show that b_o/L_F- is connected, and thus (1.11) holds. Q.E.D.

We will not need the following case, but we introduce it since it is the "untwisted" version of the preceding one.

If $\underline{A}$ and $\underline{B}$ are again small and $F: \underline{A} \longrightarrow \underline{B}$, then one has a functor

$$\underline{A}^2 = (I_{\underline{A}}, I_{\underline{A}}) \xrightarrow{\quad M_F \quad} (I_{\underline{B}}, F)$$

$$\begin{array}{ccc} A & \xrightarrow{a} & A' \\ {\scriptstyle a_1}\downarrow & & \downarrow{\scriptstyle a_2} \\ A_1 & \xrightarrow{a'} & A_1' \end{array} \qquad \begin{array}{ccc} F(A) & \xrightarrow{F(a)} & F(A') \\ {\scriptstyle F(a_1)}\downarrow & & \downarrow{\scriptstyle F(a_2)} \\ F(A_1) & \xrightarrow{F(a')} & F(A_1') \end{array}$$

<u>THEOREM 5.</u> The functor M_F is cofinal.

<u>Proof</u>. If $b_o: B \longrightarrow F(A)$ in $(I_{\underline{B}}, F)$ then $(b_o, 1_A)$: $b_o \longrightarrow M_F(1_A) = 1_{F(A)}$ which is (1.10). Let now $a_1: A_1 \longrightarrow A_1'$, $a: A \longrightarrow A_1'$ and $b_1: B \longrightarrow F(A_1)$. Then $(b_1, a): b_o \longrightarrow M_F(a_1) = F(a_1)$ is connected to $(b_o, 1_A)$ by the following cummutative diagram since $F(a).b_o = F(a_1).b_1$:

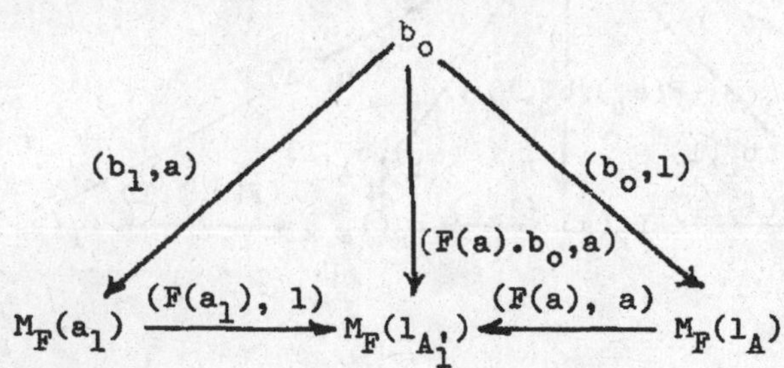

Again this shows that b_o/M_F- is connected for all $b_o \in (I_{\underline{B}}, F)$.

3. The Functor Evaluation.

Bifunctors will be used continuously in this paper, and so we recall:

DEFINITION 3. If $F: \underline{X} \times \underline{Y} \longrightarrow \underline{Z}$, then $G: \underline{Y}^* \times \underline{Z} \longrightarrow \underline{X}$ is said to be right adjoint to F, written $F \dashv G$, iff for each X, Y and Z there is a bijection of sets

$$\alpha(X, Y, Z): \underline{Z}(F(X, Y), Z) \longrightarrow \underline{X}(X, G(Y, Z))$$

natural in all variables.

The proof of the next Proposition is straightforward (see [11], Proposition 2.3, page 124, or [5]).

PROPOSITION 6. If $F: X \times Y \longrightarrow Z$ and for each Y in $\underline{Y}$, $F(\ , Y): \underline{X} \longrightarrow \underline{Z}$ has a right adjoint $G_Y: \underline{Z} \longrightarrow \underline{X}$ with adjunction α_Y, then F has a right adjoint $G: \underline{Y}^* \times \underline{Z} \longrightarrow \underline{X}$ with $G(Y, Z) = G_Y(Z)$ and $\alpha(X, Y, Z) = \alpha_Y(X, Z)$.

Now let $\underline{C}$ be a category will all small coproducts. If E is a set and C in $\underline{C}$, we let $E.C = \sum_{i \in E} C_i$ where $C_i = C$ and $\sum$ denotes the coproduct in $\underline{C}$. This gives us the functor:

$$\sum : \text{Ens} \times \underline{C} \longrightarrow \underline{C}: (E, C) \xrightarrow{(e,\, c)} (E', C') \rightsquigarrow E.C \xrightarrow{e.c} E'.C'$$

which is well known to be the left adjoint of $\underline{C}(\ , \ ): \underline{C}^* \times \underline{C} \longrightarrow \text{Ens}$ by the natural equivalence $\underline{C}(E.C, C') \cong (\underline{C}(C, C'))^E \cong \text{Ens}\,(E, \underline{C}(C, C'))$. Dually, there is a bifunctor $\Pi: \underline{C} \times \text{Ens}^* \longrightarrow \underline{C}$. When $\underline{C} = \text{Ens}$

this reduces to Ens $(E \times C, C') \cong$ Ens $(E, \text{Ens}(C, C'))$ since $E \times C \cong E.C$. Finally $\underline{C} \times \text{Ens} \cong \text{Ens} \times \underline{C} \xrightarrow{\Sigma} \underline{C}$ is the left adjoint of $\text{Ens}^* \times \underline{C} \cong \underline{C} \times \text{Ens}^* \xrightarrow{\pi} \underline{C}$ when $\underline{C}$ has all products and coproducts.

If $\underline{B}$ has all small coproducts and $\underline{A}$ is a small category we then have a functor:

$$\underline{A}^* \times \underline{A} \times \underline{B} \xrightarrow{\underline{A}(\ ,\ ) \times I} \text{Ens} \times \underline{B} \xrightarrow{\Sigma} \underline{B}$$

which, through (1.1) yields a functor

$$\begin{array}{ccc} \Gamma^b: \underline{B} \times \underline{A}^* & \longrightarrow & (\underline{A}, \underline{B}) \\ (B,A) & & \underline{A}(A,\ ).B \\ {\scriptstyle (b,a)}\downarrow & & \downarrow{\scriptstyle \underline{A}(a,1).b} \\ (B',A') & & \underline{A}(A',\ ).B \end{array}$$

where

$$\underline{A}(A,\ ).B: \underline{A} \to \underline{B}: A_1 \xrightarrow{a'} A_2 \rightsquigarrow \underline{A}(A,A_1).B \xrightarrow{\underline{A}(1,a').1} \underline{A}(A,A_2).B$$

We will also make frequent use of the functor evaluation Γ, defined by

$$\begin{array}{ccc} \Gamma: \underline{A} \times (\underline{A}, \underline{B}) & \longrightarrow & \underline{B} \\ (A, F) & & F(A) \\ {\scriptstyle (a, \eta)}\downarrow & & \downarrow{\scriptstyle F'(a)\cdot\eta(A) = \eta(A').F(a)} \\ (A', F') & & F'(A') \end{array}$$

while $\Gamma^b(\, , A)$ and $\Gamma(A, \,)$ will sometimes be written Γ^b_A and Γ_A respectively.

If $A \xrightarrow{a} A'$ is in $\underline{A}$, then $u_a: B \longrightarrow \underline{A}(A, A').B$ will designate the "canonical a-th injection" in the coproduct corresponding to a: in particular, we will write $u_1: B \longrightarrow \underline{A}(A, A).B$ for the 1-th injection where $1 = 1_A: A \longrightarrow A$. In the dual case, we will use p_a and p_1.

Now it is clear that in order to define Γ^b it suffices to impose the following restriction on $\underline{B}$:

DEFINITION 4. A category $\underline{B}$ is said to be $|\underline{A}^2|$-copowered whenever it has all copowers of the form X.B for all objects B in $\underline{B}$ and any set X of cardinality less of equal to that of the set $|\underline{A}^2|$ of arrows of $\underline{A}$.

For instance, if $\underline{B}$ is a poset with smallest element 0 and $\underline{A}$ an arbitrary small category, then $\underline{B}$ is $|\underline{A}^2|$-copowered, but does not necessarily have all coproducts. Dually, the category of finite dimensional vector spaces is $|\underline{A}^2|$-powered only if $\underline{A}$ is a finite category.

THEOREM 7. If $\underline{B}$ is $|\underline{A}^2|$-copowered, then Γ^b is the left adjoint of Γ.

Proof. For each A, B and $F: \underline{A} \longrightarrow \underline{B}$, we define a mapping $\alpha(B, A, F): (\underline{A}, \underline{B})(\Gamma^b(B, A), F) \longrightarrow \underline{B}(B, \Gamma(A, F))$ by $\alpha(B, A, F)(\Theta) = \Theta(A).u_1$ where $\Theta: \Gamma^b(B, A) \to F$, and $u_1: B \longrightarrow \underline{A}(A, A).B$ was defined above. $\alpha(B, A, F)$ is injective, for suppose

Θ': $\Gamma^b(B, A) \Rightarrow F$ with $\Theta'(A).u_1 = \Theta(A).u_1$. Then for any $A \xrightarrow{a} A'$, $\Theta(A').u_a = F(a).\Theta(A).u_1 = F(a).\Theta'(A).u_1 = \Theta'(A').u_a$ where u_a: $B \longrightarrow \underline{A}\,(\underline{A}, A').B$, and thus $\Theta(A') = \Theta'(A')$ for all A' in $\underline{A}$ which means that $\Theta = \Theta'$. If now $B \xrightarrow{b} F(A)$, then for any $A \xrightarrow{a} A'$, one has an arrow $B \longrightarrow F(A) \xrightarrow{F(a)} F(A')$ which induces a unique arrow $\underline{A}(A, A').B \xrightarrow{\Theta(A')} F(A')$ with $\Theta(A').u_a = F(a).b$. This is easily proved to be natural "in A'" and thus $\alpha(B, A, F)$ is surjective. Naturality in B, A, F is just routine and follows from the properties of the coproduct.

COROLLARY. If F: $\underline{A} \longrightarrow \underline{B}$, then every natural transformation Θ: $\Gamma^b(B, A) \Rightarrow F$ is completely determined by $\Theta(A).u_1$.

Dually, $(\underline{A}, \underline{B}) \times \underline{A} \cong \underline{A} \times (\underline{A}, \underline{B}) \xrightarrow{\Gamma} \underline{B}$, where now $\underline{B}$ is $|\underline{A}^2|$-powered, has a right adjoint $\Gamma^\#$: $\underline{A}^* \times \underline{B} \longrightarrow (\underline{A}, \underline{B})$ obtained through (1.1) from $\underline{B} \times \underline{A} \times \underline{A}^* \xrightarrow{I_{\underline{B}} \times \underline{A}^*(\ ,\ )} \underline{B} \times \mathrm{Ens}^* \xrightarrow{\pi} \underline{B}$ and every natural transformation Θ: $F \Rightarrow \Gamma^\#(A, B)$ is completely determined by $p_1.\Theta(A)$ where p_1: $B^{\underline{A}(A, A)} \longrightarrow B_1$ was defined above.

Since one of the main purposes of this article is to show the fundamental role played by Γ^b (and $\Gamma^\#$) in category theory, we now recall the fact that the Yoneda lemma is only a special case of the preceding theorem. This is done as follows (cf. also [13]):

The functor $\mathrm{Ens} \times \underline{A}^* \times \underline{A} \xrightarrow{I \times \underline{A}(\ ,\ )} \mathrm{Ens} \times \mathrm{Ens} \xrightarrow{\sum} \mathrm{Ens}$ induces through (1.1) again a functor Δ: $\mathrm{Ens} \times \underline{A}^* \longrightarrow (\underline{A}, \mathrm{Ens})$

where $\Delta(E, A)(A') = E.(\underline{A}(A, A'))$. If we let $\underline{B}$ = Ens in the definition of Γ^b, then:

PROPOSITION 8. Δ and Γ^b are naturally equivalent bifunctors.

Proof. This stems from the well known fact that if E and E' are sets, then E.E' = E'.E, coproducts being disjoint unions in Ens.

COROLLARY. (Yoneda Lemma) Let F: $\underline{A} \longrightarrow$ Ens. Then there is a bijection of sets

$$\beta(A, F): (\underline{A}, \text{Ens})(\underline{A}(A, \ ), F) \longrightarrow F(A)$$

natural in A and F, and given by $(\beta(A, F))(\Theta) = (\Theta(A))(1_A)$.

Proof. If E is the set 1 with only one element, then $1.(\underline{A}(A, \ )) = \underline{A}(A, \ )$ and Ens $(1, \Gamma(A, F)) \cong F(A)$ natural in A and F. Thus we set $\beta(A, F) = \alpha(1, A, F)$ in Theorem 7 and use the equivalence of Proposition 8. Q.E.D.

Replacing $\underline{A}$ by its dual in this section gives us dual results for contravariant functors.

4. Functorial Density

The following notion of density in Category Theory is well known: if $\underline{B}$ is a subcategory of $\underline{C}$, then $\underline{B}$ is dense in $\underline{C}$ whenever every C in $\underline{C}$ is the limit of a diagram D: $\underline{A} \longrightarrow \underline{B}$. this notion is extended in two directions in this chapter: the inclusion functor is replaced by a general functor

$\underline{B} \to \underline{C}$, and the whole thing must be "functorial" in a sense to be made precise below, and which utilizes the following notion of Kan ([5]):

DEFINITION 5. If $\underline{B}$ is any category, then $\underline{B}_d$ is the category of all diagrams over $\underline{B}$: an object is a pair $(\underline{A}, D)$ where $D: \underline{A} \to \underline{B}$, and $\underline{A}$ is small, and an arrow $(\underline{A}, D) \to (\underline{A}', D')$, with $D': \underline{A}' \to \underline{B}$, is also a pair (F, η) where $F: \underline{A} \to \underline{A}'$ and $\eta : D \to D' \circ F$. Composition is defined by $(F', \eta') \cdot (F, \eta) = (F' \circ F, (\eta' * F) \cdot \eta)$, where $D'': \underline{A}'' \to \underline{B}$, $\eta': D' \to D'' \circ F'$, and it obeys the "five rules of functorial calculus" of Godement.

It is easy to verify that every functor $G: B \to C$ induces a functor $G_d: \underline{B}_d \to \underline{C}_d$ where $G_d(\underline{A}, D) = (\underline{A}, G \circ D)$ and $G_d(F, \eta) = (F, G * \eta)$ with $(G * \eta) = G(\eta(A))$. Similarly any natural transformation $\Theta: G \to G'$, $G': \underline{B} \to \underline{C}$, induces a natural transformation $\Theta_d: G_d \to G'_d$ where $\Theta_d(D) = (1_{\underline{A}}, \Theta * D)$.

Let us assume temporarily that $\underline{C}$ is a cocomplete category. A functor $G: \underline{B} \to \underline{C}$ is then said to be <u>functorially codense</u> whenever there exists a (not necessarily unique) functor $G': G \to \underline{B}_d$ such that the composition:

$$(4.13) \qquad \underline{C} \xrightarrow{G'} \underline{B}_d \xrightarrow{G_d} \underline{C}_d \xrightarrow{\text{Colim}} \underline{C}$$

is naturally equivalent to the identity endofunctor of $\underline{C}$, where Colim is the functor associating to each diagram $D: \underline{A} \to \underline{C}$ its colimit in $\underline{C}$ and to each morphism $(F, \eta): D \to D'$ where $F: \underline{A} \to \underline{A}'$, $D': \underline{A}' \to \underline{C}$, $\eta : D \to D' \circ F$, the unique arrow

Colim (D) $\rightarrow$ Colim (D') induced in $\underline{C}$ by η. (i.e., D'$\bullet$F is a "subdiagram" of D). But it is quite clear that it is not necessary to assume that $\underline{C}$ has all colimits. In fact, it suffices to assume that $\underline{C}$ has all $(G_d \bullet G')(C) = G\bullet(G'(C))$ - colimits, where $G'(C)$: $\underline{A} \rightarrow \underline{B}$, $\underline{A}$ being some small category: then, "write everything in words". This is what we will use as a definition of functorial codensity: i.e., we will only assume that $\underline{C}$ is $G\bullet(G'(\underline{C}))$-cocomplete, but still, for commodity, we will keep the notation of (4.13) above. Of course, the dual will be called functorial density, (cf. [6] and [12], (2.27), for "non functorial" density).

As immediate examples of functorially codense functors we have the usual embedding of the rationals into the reals and the functor $\underline{1} \rightarrow$ Ens sending the object of 1 into the set 1 with only one element. We will give others below, but first we recall that the choice of G' in the definition of functorial codensity may not be unique:
For instance, the identity functor I: $\underline{A} \rightarrow \underline{A}$ is left adequate (Proposition 12) and thus any object A is the colimit of the obvious functor $\underline{A}/A \rightarrow \underline{A}$ (cf. remarks after corollary 1 of the next theorem). But A is also the colimit of $\underline{1} \rightarrow \underline{A}$: $1 \rightsquigarrow A$, and thus there are two possible choices for I'.

THEOREM 9. If $\underline{B}$ is $|\underline{A}^2|$-copowered then Γ^b: $\underline{B}\times\underline{A}^* \rightarrow (\underline{A}, \underline{B})$ is functorially codense. In fact, if F: $\underline{A} \rightarrow \underline{B}$, we choose $(\Gamma^b)'(F)$ to be the following diagram in $\underline{B}\times\underline{A}^*$:

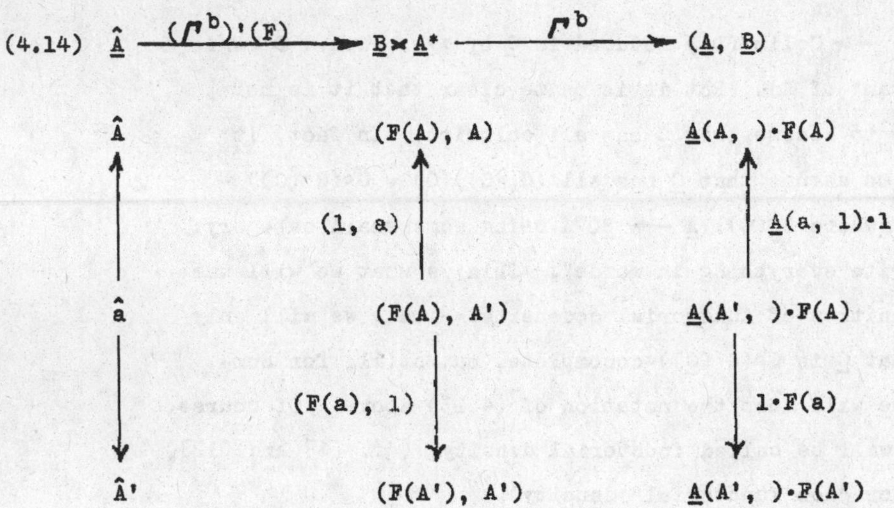

with $A \xrightarrow{a} A'$ in $\underline{A}$, and if $\eta: F \to F'$, $F': \underline{A} \to \underline{B}$, then $(\Gamma^b)'(\eta) = (I, \bar{\eta})$ where $I: \hat{\underline{A}} \to \hat{\underline{A}}$ and $\bar{\eta}: (\Gamma^b)'(F) \to (\Gamma^b)'(F')$ is defined by $\bar{\eta}(\hat{a}) = (\eta(A), 1_{A'})$. Thus F is the colimit of (4.14).

<u>Proof</u>. Using the same α as in the Proof of Theorem 7, define $\Theta: \Gamma^b \circ (\Gamma^b)'(F) \to F_{\hat{\underline{A}}}$, where $F_{\hat{\underline{A}}}: \hat{\underline{A}} \to (\underline{A}, \underline{B}): \hat{a} \rightsquigarrow F$ is the constant diagram, by $\Theta(\hat{a}) = (\alpha(F(A), A', F))^{-1}(F(a)): \underline{A}(A', \ )\cdot F(A) \to F$, for all $A \xrightarrow{a} A'$.

The naturality and universality of Θ are a direct consequence of the adjunction relation $\Gamma^b \dashv \Gamma$.

COROLLARY 1. If $\underline{B}$ is $|\underline{A}^2|$-copowered and $F: \underline{A} \to \underline{B}$, then for any A in $\underline{A}$, F(A) is the colimit of the functor:

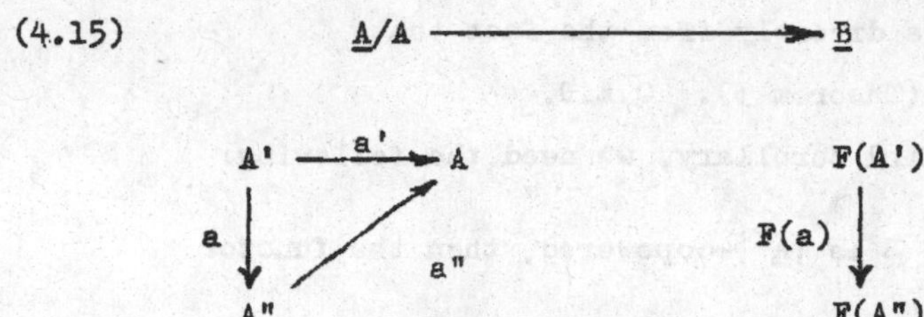

<u>Proof</u>. See the Corollary of Proposition 14 in the next section. Q.E.D.

The preceeding result is of course "functorial in A", it is true for any category $\underline{B}$ since it can easily be established directly without using Theorem 9 and by taking F to be identity endofunctor of $\underline{A}$, an even more trivial result is obtained (see remarks before Theorem 9). As an application consider two posets $\underline{A}$ and $\underline{B}$, $\underline{B}$ with smallest element 0: the functors $\Gamma^b(B, A)$ are then those of the form $f_{b,\, a}$ where for all x in $\underline{A}$, $f_{b,\, a}(x) = b$ if $a \leq x$, and 0 otherwise; Theorem 9 then says that every monotone mapping $f: \underline{A} \longrightarrow \underline{B}$ is the sup of the set $\{f_{f(a),\, a} \mid a \in A\}$.

COROLLARY 2. If $\underline{B}$ is $|\underline{A}^2|$-copowered, then every functor $F: \underline{A} \longrightarrow \underline{B}$ is the colimit of:

(4.16) $$(\tilde{\underline{A}})^* \longrightarrow (\underline{A}, \underline{B})$$

$$\begin{array}{ccc} A & \xrightarrow{a} & A' \\ {\scriptstyle a_1}\downarrow & & \uparrow{\scriptstyle a_2} \\ A_1 & \xrightarrow{a'} & A_1' \end{array} \qquad\qquad \begin{array}{c} \underline{A}(A', \ )\cdot F(A) \\ \downarrow{\scriptstyle \underline{A}(a_2,\ )\cdot F(a_1)} \\ \underline{A}(A_1', \ )\cdot F(A_1) \end{array}$$

Proof. This follows directly from the fact that $T_0 : \hat{A} \longrightarrow (\tilde{A})^*$ is cofinal (Theorem 3). Q.E.D.

Before stating a third corollary, we need the following:

PROPOSITION 10. If $\underline{B}$ is $|\underline{A}^2|$-copowered, then the functor

$$(4.17) \qquad \Gamma^b\_/F = (\Gamma^b, \epsilon_F) \longrightarrow \underline{L}^*_F$$

$$\begin{array}{ccc} \underline{A}(A, \;)\cdot B & \xrightarrow{\Theta} & F \\ \downarrow \underline{A}(a, \;)\cdot b_1 & \nearrow \Theta' & \\ \underline{A}(A', \;)\cdot B' & & \end{array} \qquad \begin{array}{ccc} B & \xrightarrow{\Theta(A)\cdot u_1} & F(A) \\ b_1 \downarrow & & \uparrow F(a) \\ B' & \xrightarrow{\Theta'(A')\cdot u_1} & F(A') \end{array}$$

is an isomorphism, where $A' \xrightarrow{a} A$.

Proof. This is simply a restatement of Theorem 7.

COROLLARY 3. If moreover $\underline{B}$ is small, then $F: \underline{A} \longrightarrow \underline{B}$ is the colimit of:

$$(4.18) \qquad \underline{L}^*_F \xrightarrow{L} \underline{B} \ltimes \underline{A}^* \xrightarrow{\Gamma^b} (\underline{A}, \underline{B})$$

$$\begin{array}{ccc} B & \xrightarrow{b} & F(A) \\ b_1 \downarrow & & \uparrow F(a) \\ B' & \xrightarrow{b'} & F(A') \end{array} \qquad \begin{array}{c} (B, A) \\ \downarrow (b_1, a) \\ (B', A') \end{array} \qquad \begin{array}{c} \underline{A}(A, \;)\cdot B \\ \downarrow \underline{A}(a, \;)\cdot b_1 \\ \underline{A}(A', \;)\cdot B' \end{array}$$

Proof. From Theorem 4, Corollary 2 above and the fact that $\underline{B}$ small implies that $\underline{L}^*_F$ is also small. Q.E.D.

Those results are far from new: Corollary 2 is implicit in ([1]) (see section 8), Theorem 9 can be obtained from ([5]) as we will see later, and other people have used them, in particular Ulmer ([12]). But we meant to show that they can all be derived from the fact that Γ has a left adjoint. Of course by using $\Gamma^{\#}$, limits over $(\hat{A})^*$, $\tilde{A}$ and $(\in_F, \Gamma^{\#}) \cong (\in_1, \underline{B}(F(\ ),\ )$ we get dual results, and there will be corresponding versions for contravariant functors.

The next proposition can be shown to be a consequence of Theorem 9 using a general result of [13] and the fact that $\underline{1} \rightsquigarrow \mathrm{Ens}: 1 \rightsquigarrow 1$ is left adequate (Definition 6), but we sketch a direct proof (see also remarks after Proposition 26).

PROPOSITION 11. The co-Yoneda functor $Y^*: \underline{A}^* \longrightarrow (\underline{A}, \mathrm{Ens})$ is functorially codense. In fact, if $F: \underline{A} \longrightarrow \mathrm{Ens}$, then $(Y^*)'(F)$ is the following diagram in $\underline{A}^*$:

(4.19) $$(A, \mathrm{Ens})_r/F \xrightarrow{(Y^*)'(F)} \underline{A}^* \xrightarrow{Y^*} (\underline{A}, \mathrm{Ens})$$

$$\begin{array}{ccc}
\begin{array}{ccc} \underline{A}(A,\ ) & \xrightarrow{\Theta_x} & F \\ {\scriptstyle \underline{A}(a,\ )}\uparrow & \vdots\uparrow & \nearrow{\scriptstyle \Theta_{x'}} \\ \underline{A}(A',\ ) & & \end{array}
&
\begin{array}{c} A \\ \downarrow{\scriptstyle a} \\ A' \end{array}
&
\begin{array}{c} \underline{A}(A,\ ) \\ {\scriptstyle \underline{A}(a,\ )}\uparrow \\ \underline{A}(A',\ ) \end{array}
\end{array}$$

where Θ_x corresponds to $x \in F(A)$ by the Yoneda lemma and if $\eta: F \Rightarrow F'$, then $(Y^*)'(\eta) = (Y^*_\eta, 1)$ where $Y^*_\eta(\Theta_x) = \Theta_{(\eta(A))(x)}$: $\underline{A}(A,\ ) \Rightarrow F'$ corresponds to $(\eta(A))(x) \in F'(A)$ by the Yoneda

Lemma again, and $Y^*_\eta(\underline{A}(a, \ )) = \underline{A}(a, \ )$, while $1(\Theta_x)$: $((Y^*)'(F))(\Theta_x) = A \longrightarrow A = ((Y^*)'(F'))\cdot Y^*_\eta(\Theta_x)$ is the identity. Thus F is the colimit of (4.19).

Proof: From Theorem 9, F is the colimit of $\hat{A} \xrightarrow{(\Gamma^b)'(F)} \mathrm{Ens} \times \underline{A}^* \xrightarrow{\Gamma^b} (\underline{A}, \mathrm{Ens})$ and the result follows from the fact that each set $\{\Theta_x : \underline{A}(A, \ ) \to F \mid x \in F(A)\}$, for each A, generates a unique natural transformation $F(A)\cdot A(A, \ ) \cong \underline{A}(A, \ )\cdot F(A) \xrightarrow{\Theta(\hat{A})} F$ (by Proposition 8) which composed with the canonical coproduct $\underline{A}(A, \ ) \xrightarrow{u_x} F(A)\cdot\underline{A}(A, \ )$ is equal to Θ_x. (Or apply Proposition 14 to $(\underline{A}, \mathrm{Ens}) \xleftarrow{Y} A^* \xrightarrow{Y} (\underline{A}, \mathrm{Ens})$) Q.E.D.

This theorem is better known in the following version, which probably goes back to Gabriel:

COROLLARY. Every functor $F: \underline{A} \longrightarrow \mathrm{Ens}$, $\underline{A}$ small, is the colimit of representable functors.

That result of course holds for contravariant functors by taking $\underline{A}^*$ instead of $\underline{A}$, but there is no "limit" conterpart besides the trivial one $Y^*: \underline{A} \longrightarrow (\underline{A}, \mathrm{Ens})^*$. From (1.3) we could also use the category of F-pointed objects instead of $(\underline{A}, \mathrm{Ens})_r/F$.

DEFINITION 6. (Isbell) If $F: \underline{A} \longrightarrow \underline{B}$, then F generates $\underline{B}$ whenever $\overline{F}: \underline{B} \longrightarrow (\underline{A}^*, \mathrm{Ens})$, (cf. (1.4)), is an embedding, and F is left adequate if moreover $\overline{F}$ is full (in fact, $\overline{F}$ is always faithfull on objects), $\underline{A}$ always small.

Equivalently, F is left adequate iff there is a bijection of sets $\mathrm{Nat}(\underline{B}(F(\ ), B), \underline{B}(F(\ ), B')) \cong \underline{B}(B, B')$: $\underline{B}(1, b) \rightsquigarrow b$

(cf. [4] and [6]):

PROPOSITION 12. If $F\colon \underline{A} \to \underline{B}$ is left adequate then it is functorially codense. In fact, for any B in $\underline{B}$, $F'(B)$ is the following diagram in $\underline{A}$:

(4.20)

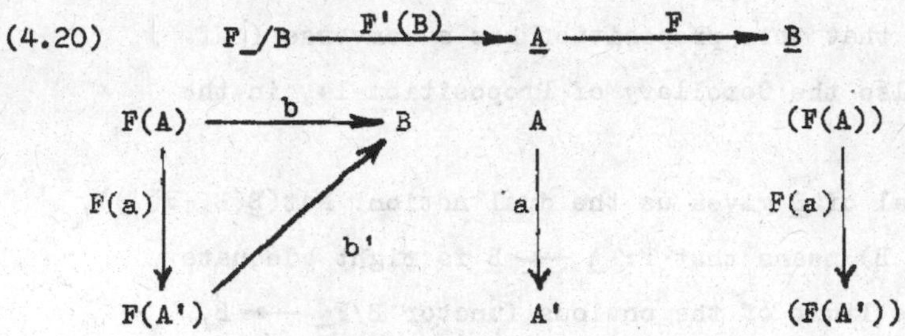

and if $b''\colon B \to B'$, then $F'(b'') = (F_{b''}, 1)$ where $F_{b''}\colon F\_/B \to F\_/B'$: $b \xrightarrow{F(a)} b' \rightsquigarrow b''b \xrightarrow{F(a)} b''b'$. Thus B is the colimit of (4.20).

Proof. We already noted in (1.4), that $F\_/B \cong (\underline{A}^*, \mathrm{Ens})_r/\overline{F}(B)$, and using Proposition 11, $\overline{F}(B)\colon A^* \to \mathrm{Ens}$ is the colimit of

$$F\_/B \longrightarrow (\underline{A}^*, \mathrm{Ens})_r/\overline{F}(B) \xrightarrow{Y'(\overline{F}(B))} \underline{A} \xrightarrow{Y} (\underline{A}^*, \mathrm{Ens})$$

$$\begin{array}{ccccccc} FA \xrightarrow{b} B & & \underline{A}(\ ,A) \xrightarrow{\Theta_b} \underline{B}(F(\ ),B) & & A & & \underline{A}(\ ,A) \\ F(a)\downarrow \nearrow b' & & \underline{A}(\ ,a)\downarrow \nearrow \Theta_{b'} & & a\downarrow & & \downarrow \underline{A}(1,a) \\ FA' & & \underline{A}(\ ,A) & & A & & \underline{A}(\ ,A) \end{array}$$

Now B is the colimit of (4.20) since there is a 1-1 correspondence between the arrows b and Θ_b, by the co-Yoneda lemma, and also between the arrows $B \to B'$ and $\underline{B}(F(\ ),B) \to \underline{B}(F(\ ),B')$.

since F is adequate. Also every $B \xrightarrow{b''} B'$ induces $\eta = \underline{B}(1, b'')$: $\overline{F}(B) \to \overline{F}(B')$ and $Y_{\eta}(\theta_b) = \theta_{(\eta(A))(b)} = \theta_{b''.b}$ which gives $F_{b''}$ defined as above, and the rest is obvious. Q.E.D.

Actually, Lambek uses the category $(\underline{A}^*, \text{Ens})_r/\overline{F}(B)$ instead of $F\_/B$, but as we just saw above, the result is the same. He also shows that this proposition has a converse (c.f. [12], (1.7)). See also the Corollary of Proposition 14, in the next section.

Taking the dual of $\underline{B}$ gives us the dual notion: $\text{Nat}(\underline{B}(B, F(\ )), \underline{B}(B', F(\ ))) \cong \underline{B}(B', B)$ means that $F: \underline{A} \to \underline{B}$ is right adequate and any B is then the limit of the obvious functor $B/F\_ \to B$. Moreover the Yoneda lemma implies that the Yoneda functor is left adequate and thus Proposition 11 follows from the above result. The functor $\underline{1} \to \text{Ens}: 1 \rightsquigarrow 1$ and the embedding of the rationals into the reals are also left adequate and thus functorially codense as we already noted just before Theorem 9.

Is the same true for Γ^b and $\Gamma^{\#}$? Almost:

THEOREM 13. If $\underline{A}$ and $\underline{B}$ are small, then $\Gamma^b: \underline{B} \times \underline{A}^* \to (\underline{A}, \underline{B})$ is left adequate if $\underline{B}$ is $|\underline{A}^2|$-copowered.

<u>Proof</u>. Theorem 7 means that $(\underline{A}, \underline{B})(\Gamma^b(\ ,\ ), F)$, for any $F: \underline{A} \to \underline{B}$, is naturally equivalent to $\underline{B}(\ , F(\ ))$, which yields $\overline{F}^*: \underline{B}^* \to (\underline{A}, \text{Ens})$ through (1.1). Thus we must show that:

$$(4.21)\quad (\underline{A}, \underline{B}) \xrightarrow{\overline{\Gamma^b}} ((\underline{B} \times \underline{A}^*)^*, \text{Ens}) \cong (\underline{B}^* \times \underline{A}, \text{Ens}): F \rightsquigarrow \underline{B}(\ , F(\ ))$$

is a full embedding, or equivalently, that the correspondence:

$\mathrm{Nat}\,(\underline{B}(\ ,F(\ )),\underline{B}(\ ,F'(\ )))\cong \mathrm{Nat}\,(F,F'):\underline{B}(\ ,\eta(\ ))\leftrightsquigarrow\eta$

is a natural bijection. But this follows directly from the co-Yoneda lemma. Q.E.D.

Bénabou defines a profunctor from $\underline{A}$ to $\underline{B}$ as a functor $\underline{A}^*\times\underline{B}\longrightarrow \mathrm{Ens}$. Thus Theorem 13 means that $F\rightsquigarrow(\underline{A}^*\times B)^*\cong \underline{B}^*\times\underline{A}\xrightarrow{\underline{B}(\ ,F(\ ))}\mathrm{Ens}$ gives a full embedding of the category $(\underline{A},\underline{B})$ into that of "contravariant profunctors" from $\underline{A}$ to $\underline{B}$. Also applying Proposition 12 to Γ^b gives the fact that any $F:\underline{A}\longrightarrow\underline{B}$ is the colimit of:

$$L^*_F\simeq\Gamma^b/F\longrightarrow\underline{B}\times\underline{A}^*\xrightarrow{\Gamma^b}(\underline{A},\underline{B})$$

$$B\longrightarrow F(A)\rightsquigarrow\underline{A}(A,\ )\cdot B\longrightarrow F\rightsquigarrow(B,A)\rightsquigarrow\underline{A}(A,\ )\cdot B$$

which is precisely Corollary 3 of Theorem 9 (using (4.17)). (See Remark 2 in Section 7.)

Finally in [12], Ulmer calls a functor $F:\underline{A}\longrightarrow\underline{B}$ dense, $\underline{A}$ not necessarily small, whenever the result of Proposition 12 holds (i.e., every B is the colimit of (4.20)). But he must allow "large diagrams" and by the converse of Proposition 12 (see [6]) dense = left adequate when $\underline{A}$ is small: thus, all the functors considered so far, including Γ^b, are dense in Ulmer's sense.

5. Kan Extensions and Globalisation

Let $\underline{A}$ be small and $\underline{B}\xleftarrow{F}\underline{A}\xrightarrow{H}\underline{C}$ satisfy the following two conditions:

(5.22)· $\underline{C}$ has all coproducts of the form $\underline{B}(F(A'), B) \cdot H(A)$ for all A, A' and B.

(5.23) The colimit of $\hat{\underline{A}} \to \underline{C}$: $\hat{a} \rightsquigarrow \underline{B}(F(A'), B) \cdot H(A)$, where $A \xrightarrow{a} A'$ in $\underline{A}$, exists for all B in $\underline{B}$ where the latter is essentially the functor $\hat{\underline{A}} \xrightarrow{(\Gamma^b)'(H)} \underline{C} \times \underline{A}^* \xrightarrow{1 \times F^*_*} \underline{C} \times \underline{B}^* \xrightarrow{\Gamma^b} (\underline{B}, \underline{C})$ when $\underline{B}$ is small and $\underline{C}$ is $|\underline{B}^2|$-copowered.

If (5.22) and (5.23) hold, we define $(K_F(H))(B)$ to be the colimit of (5.23), which gives us a functor $K_F(H)$: $\underline{B} \to \underline{C}$.

When $\underline{B}$ is small and $\underline{C}$ is $(\underline{B}^2)$-copowered then $K_F(H)$ is equal to colim $(\Gamma^b \circ (1 \times F^*_*) \circ (\Gamma^b)'(H))$ which in turn gives a functor K_F: $(\underline{A}, \underline{C}) \to (\underline{B}, \underline{C})$ where, if $\eta: H \to H'$ in $(\underline{A}, \underline{C})$, then $K_F(\eta)$ = colim $(\Gamma^b \circ (1 \times F^*_*)_* \bar{\eta})$, $\bar{\eta}$ being defined by $\bar{\eta}(\hat{a}) = (\eta(A), 1_{A'})$.

If conversely $K_F(H)$ = colim $(\Gamma^b \circ (1 \times F^*_*) \circ (\Gamma^b)'(H))$ exists and $\underline{C}$ is $|\underline{B}^2|$-copowered, then for all B, Γ_B is the left adjoint of Γ^b_B and thus preserves colimits, and we get the above definition for $(K_F(H)(B)$ back.

In particular $K_F(H)$ exists when $\underline{C}$ is $|\underline{B}^2|$-copowered and is $\hat{\underline{A}}$-cocomplete, and if large diagrams are allowed (as in [12]) then $\underline{A}$ need not be small in order that $K_F(H)$ be a functor. $K_F(H)$ is in fact the (right) Kan F-extension (or Kan F-coextension) of H. For the dual case use $(A)^*$, $\Gamma^{\#}$ and limits, and for the universal property of $K_F(H)$, and in fact for its definition, see the discussion after Proposition 30.

Keeping the same notation as above, we will need the following for future references:

(5.24) $\underline{C}$ has all copowers of the form $\underline{B}(F(A'), B)\cdot C$ for all A', B and C.

(5.25) The colimit of $\hat{\underline{A}} \longrightarrow (\underline{B}, \underline{C})$: $\hat{a} \rightsquigarrow \underline{B}(F(A'), \ )\cdot H(A)$ exists.

(5.24) implies (5.22) and (5.23) is equivalent to (5.25) when $\underline{C}$ is $|\underline{B}^2|$-copowered.

PROPOSITION 14. If $\underline{A}$ and $\underline{B}$ are small and $\underline{B} \xleftarrow{F} \underline{A} \xrightarrow{H} \underline{C}$ satisfy (5.22) and (5.23), then for any B in $\underline{B}$, $(K_F(H))(B)$ is the colimit of the functor $F\_/B \xrightarrow{F'(B)} \underline{A} \xrightarrow{H} \underline{C}$: $F(A) \xrightarrow{b} B \rightsquigarrow A \rightsquigarrow (H(A))$, while for any $B \xrightarrow{b''} B'$, $(K_F(H))(b'') = \text{colim } (H_*F'(b''))$ and if $\eta : H \Rightarrow H'$, $K_F(\eta)$ is defined by $(K_F(\eta))(B) = \text{colim } (\eta_*F'(B))$, for every B in $\underline{B}$. (c.f. Proposition 12 for $F'(b'')$).

Proof. $F\_/B$ is small since $\underline{B}$ is. Now let η: $H\circ(F'(B)) \Rightarrow C_{F\_/B}$ (cf. Theorem 2 for the latter), C in $\underline{C}$, i.e., for each $F(A) \xrightarrow{b} B$, we have an arrow $\eta(b)$: $(H(A)) \longrightarrow C$, and if $A \xrightarrow{a} A'$ then $\eta(b')\cdot H(a) = \eta(b)$ where $F(A') \xrightarrow{b'} B$. This generates a natural transformation $\bar{\eta}$: $\Gamma_B\circ\Gamma^b\circ(1\times F^*_*)\circ((\Gamma^b)'(H)) \Rightarrow C_{F\_/B}$ where $\bar{\eta}(\hat{A})\cdot u_b = \eta(b)$ for all b, where u_b: $H(A) \longrightarrow \underline{B}(F(A),B)\cdot H(A)$ was defined just before definition 4. The converse is also true: given $\bar{\eta}$, define η by $\eta(b) = \bar{\eta}(\hat{A})\cdot u_b$. The rest of the proof is straightforward. Q.E.D.

COROLLARY. Let $\underline{B} \xleftarrow{F} \underline{A} \xrightarrow{H} \underline{C}$ with $\underline{A}$ and $\underline{B}$ small.

(1) If $\underline{C}$ is $|\underline{A}^2|$-copowered and (5.24) and (5.23) both hold for all H: $\underline{A} \longrightarrow \underline{C}$, then K_F is the Γ^b-(where Γ^b: $\underline{C}\times\underline{A}^* \longrightarrow (\underline{A}, \underline{C})$) Kan extension of $\underline{C}\times\underline{A}^* \longrightarrow (\underline{B}, \underline{C})$: $(C, A) \rightsquigarrow \underline{B}(F(A'), \ )\cdot C$ (the

latter being equal to $\underline{C} \times \underline{A}^* \xrightarrow{1 \times F^*_*} \underline{C} \times \underline{B}^* \xrightarrow{\Gamma^b} (\underline{B}, \underline{C})$ when $\underline{C}$ is $|\underline{B}^2|$-copowered). This shall henceforth be written $K_F = K_{\Gamma^b}(\Gamma^b \bullet (1 \times F^*_*))$ even when $\underline{C}$ only satisfies (5.24).

(2) If (5.22) and (5.23) hold then $K_F(H) \cong (K_Y(H)) \bullet \overline{F}$ (see (1.4) for $\overline{F}$).

(3) If $\underline{A} = \underline{B}$, F denotes the identity endofunctor and $\underline{C}$ is $|\underline{A}^2|$-copowered, then for any A in $\underline{A}$, H(A) is the colimit of $\underline{A}/A \longrightarrow \underline{A} \xrightarrow{H} \underline{C}$: $A' \longrightarrow A \rightsquigarrow H(A')$. (This is corollary 1 of Theorem 9).

<u>Proof</u>. (1) From Proposition 14, $(K_{\Gamma^b}(\Gamma^b \bullet (1 \times F^*_*))(H)$ is the colimit of (essentially) the functor $\underline{L}^*_H \cong$
$\Gamma^b\text{-}/H \longrightarrow \underline{C} \times \underline{A}^* \xrightarrow{1 \times F^*_*} \underline{C} \times \underline{B}^* \xrightarrow{\Gamma^b} (\underline{B}, \underline{C})$: $C \longrightarrow H(A) \rightsquigarrow (\underline{B}(F(A), \ )) \cdot C$
But this is equal to $K_F(H)$ by the cofinality of $\hat{\underline{A}} \xrightarrow{To} (\hat{\underline{A}})^* \xrightarrow{L_F} \underline{L}^*_H$.

(2) For any B in $\underline{B}$, $(K_Y(H))(\overline{F}(B))$ is the colimit of
$Y\text{-}/\underline{B}(F(\ ), B) \longrightarrow \underline{A} \xrightarrow{H} \underline{C}$: $\underline{A}(\ , A) \longrightarrow \underline{B}(F(\ ), B) \rightsquigarrow H(A)$ by Proposition 14 but $Y\text{-}/(F(\ ), B) \cong F\text{-}/B$ by (1.4).

(3) For any A in $\underline{A}$, Γ_A is the left adjoint of Γ^b_A and thus commutes with colimits. So $H(A) = \text{colim}\ (\hat{\underline{A}} \xrightarrow{(\Gamma^b)'(H)} \underline{C} \times \underline{A}^* \xrightarrow{\Gamma^b} (\underline{A}, \underline{C}) \xrightarrow{\Gamma_A} \underline{C})$ by Theorem 9, and by Proposition 14 this is equal to the colimit of $F\text{-}/A = \underline{A} \longrightarrow \underline{A} \xrightarrow{H} \underline{C}$.

We cannot of course expect in general to have $(K_F(H)) \bullet F \cong H$. For instance, if $F: \underline{1} \longrightarrow \text{Ens}: 1 \rightsquigarrow E$, $H: \underline{1} \longrightarrow \text{Ab}: 1 \rightsquigarrow A$, A being a non trivial abelian group, then $(K_F(H))(E) = \bigoplus A_e$, $A_e \cong A$ and e varies over the set of endomaps of E, where E is in fact any set containing more than one element, and $\bigoplus$ denotes the direct sum in the category of abelian groups (cf. also [14]).

If now $G: \underline{B} \to \underline{C}$ is functorially codense, $\underline{B}$ and $\underline{C}$ small and $H: \underline{B} \to \underline{E}$, $\underline{E}$ any $G\circ(G'(\underline{C}))$-cocomplete category, then we can define a functor $\text{cog}_{G'}: (\underline{B}, \underline{E}) \to (\underline{C}, \underline{E})$:

$H \xrightarrow{\eta} H' \rightsquigarrow \text{Colim}\circ H_d\circ G' \xrightarrow{\text{colim}\cdot \eta_d \cdot G'} \text{Colim}\circ H'_d\circ G'$ (see Definition 5 for H_d and η_d). In fact, $(\text{cog}_{G'}(H))(C) = \text{colim}\,(H\circ G'(C))$ and if $c: C \to C'$ in $\underline{C}$, $\text{cog}_{G'}(c) = \text{colim}\,(H_*(G'(c)))\cdot \text{Cog}_{G'}$ is the functor co-globalisation (with respect to G'), and we will also use the above terminology even when $\underline{B}$ and $\underline{C}$ are not small categories: in that case, $\text{cog}_{G'}$ is only a "metafunctor" but $\text{cog}_{G'}(H)$ remains a valid functor. The dual will be called $\text{glo}_{G'}$ for globalisation. The next theorem shows the relation between globalisation and Kan extensions.

THEOREM 15. If $\underline{A}$ and $\underline{B}$ are small, $F: \underline{A} \to \underline{B}$ left adequate and $\underline{C}$ satisfies (5.22) and (5.23) for all $H: \underline{A} \to \underline{C}$, then there is a natural equivalence $\text{cog}_{F'} \cong K_F: (\underline{A}, \underline{C}) \to (\underline{B}, \underline{C})$, where F' was defined in Proposition 12. If $\underline{B}$ is not small, then $\text{cog}_{F'}(H) \cong K_F(H)$ for all H.

Proof. For any $H: \underline{A} \to \underline{C}$ and B in $\underline{B}$, $(\text{cog}_{F'}(H))(B) = \text{colim}\,(H\circ(F'(B)) = (K_F(H))(B)$ and same for the arrows, using Proposition 14.

COROLLARY. (1) $\text{cog}_{Y'}(H) \cong K_Y(H)$ for all $H: \underline{A} \to \underline{C}$.

(2) If $\underline{B}$ is $|\underline{A}^2|$-copowered and $\underline{C}$ is $G\circ((\Gamma^b)'(F))$-cocomplete for all $F: \underline{A} \to \underline{B}$, then $(\text{cog}_{(\Gamma^b)'}(G))(F) = \text{colim}\,(G\circ((\Gamma^b)'(F))$. If $\underline{B}$ is also small then $\text{cog}_{(\Gamma^b)'} \cong K_{\Gamma^b}$.

Nota Bene. For simplicity we will keep the notation $K_{\Gamma^b}(G)$ even when $\underline{B}$ is not small and write $K_{\Gamma^\#}$ for $glo_{(\Gamma^\#)'}$.

DEFINITION 7. Let $\underline{B}$ and $\underline{C}$ be small and $G: \underline{B} \longrightarrow \underline{C}$ be functorially codense. If $\underline{E}$ is any other category, then $(\underline{C}, \underline{E})_{G'}$ is the full subcategory of $(\underline{C}, \underline{E})$ consisting of all functors preserving the colimit (cf. [11] page 52) of all diagrams of the form $G \circ G'(C)$ for all C in $\underline{C}$. The restriction of (G, I) to $(\underline{C}, \underline{E})_{G'}$ is then denoted $(G, I)'$.

THEOREM 16. The functor

$$(\underline{C}, \underline{E})_{G'} \xrightarrow{(G, I)'} (\underline{B}, \underline{E}) \xrightarrow{cog_{G'}} (\underline{C}, \underline{E})$$

where G is functorially codense and E has all $H \circ G \circ G'(C)$-colimits, for all $H: \underline{C} \longrightarrow \underline{E}$ and C in $\underline{C}$ is naturally equivalent to the canonical embedding of $(\underline{C}, \underline{E})_{G'}$ into $(\underline{C}, \underline{E})$.

Proof: $(cog_{G'} \circ (G, I)')(H) = Colim \circ H_d \circ G_d \circ G'$, and for each C in $\underline{C}$, this is equal to colim $(H \circ G \circ G'(C))$, which is $H(\text{colim } (G \circ G'(C)))$ by the choice of H, while colim $(G \circ G'(C)) \cong C$, all the isomorphisms being natural. The rest follows easily.

COROLLARY. If $\underline{B}$ is $|\underline{A}^2|$-copowered and $H: (\underline{A}, \underline{B}) \longrightarrow \underline{C}$ preserves $\Gamma^b \circ (\Gamma^b)'(F)$-colimits then $\underline{C}$ is $H \circ \Gamma^b \circ (\Gamma^b)'(F)$-cocomplete for all $F: \underline{A} \longrightarrow \underline{B}$ and $K_{\Gamma^b}(H \circ \Gamma^b) \cong H$.

THEOREM 17. If $\underline{A}$ and $\underline{B}$ are small, $\underline{C}$ satisfies (5.22) and (5.23) for all $H: \underline{A} \longrightarrow \underline{C}$, and $F: \underline{A} \longrightarrow \underline{B}$ is a left adequate full embedding, then the functor $(\underline{A}, \underline{C}) \xrightarrow{K_F} (\underline{B}, \underline{C}) \xrightarrow{(F, I)} (\underline{A}, \underline{C})$ is naturally equivalent to the identity. Thus $(K_F(H))\circ F \cong H$.

Proof. For any $H: \underline{A} \to \underline{C}$, $(((F, I)\circ K_F)(H))(A) =$ $\mathrm{colim}(H\circ F'(F(A)))$ where $H\circ F'(F(A)): F(A') \xrightarrow{b} F(A) \rightsquigarrow A' \rightsquigarrow H(A')$ for all A (cf. (4.20)). But by the conditions imposed on F, there is a unique $A' \xrightarrow{a} A$ such that $b = F(a)$. From this, it is easy to see that the above colimit is indeed H(A). The same holds for the arrows.

COROLLARY. If $H: (\underline{A}^*, Ens) \to \underline{C}$ preserves $Y\circ Y'(F)$-colimits (c.f. (4.19)) and $\underline{C}$ is $H\circ Y\circ Y'(F)$-cocomplete for all $f: \underline{A}^* \to Ens$, then $K_Y(H\circ Y) \cong H$. If $D: \underline{A} \to \underline{C}$ and $\underline{C}$ is $D\circ Y'(F)$-cocomplete for all $F: \underline{A}^* \to Ens$ then $(K_Y(D))\circ Y \cong D$.

In the above two theorems, one could drop the "smallness" restriction by using "metafunctors" again.

6. Some Adjoint Functor Theorems

From the next result we intend to derive the theorems of André [1] and Kan [5]. Although this result could be proved from a theorem announced by Ulmer as we will show afterwards, we give a direct proof of it due to its importance in this article.

THEOREM 18. Let $D: \underline{B}\times\underline{I} \to \underline{C}$ be functorially codense, with a right adjoint $D^{\#}: \underline{I}^*\times\underline{C} \to \underline{B}$ where $\underline{B}$ and $\underline{I}$ are arbitrary categories. If now $G: \underline{B}\times\underline{I} \to \underline{E}$ and $H: \underline{E} \to \underline{C}$ are such that $G \dashv D^{\#}\circ(I_{\underline{I}}{}^* \times H)$, where $\underline{E}$ is $G\circ(D'(\underline{C}))$-cocomplete, and $I_{\underline{I}}{}^*$ is the identity endofunctor of $\underline{I}^*$, then $\mathrm{cog}_{D'}(G) \dashv H$.

Proof. In view of Proposition 6, $G \dashv D^{\#} \circ (I_{\underline{I}}^{*} \times H)$ could be written: for all I in $\underline{I}$, $G(\ , I) \dashv D^{\#}(I, \ ) \circ H$. Now let $G = D^{\#} \circ (I_{\underline{I}}^{*} \times H)$ and $\theta: G \dashv G^{\#}$ and $\eta: D \dashv D^{\#}$ be the adjunctions. Then for each B, I, E we have bijections $\underline{E}(G(B, I), E) \xrightarrow{\theta(B, I, E)} \underline{B}(B, G^{\#}(I, E)) = \underline{B}(B, D^{\#}(I, H(E)) \xrightarrow{\eta^{-1}(B, I, H(E))} \underline{C}(D(B, I), H(E))$ natural in all variables. Now for any C in $\underline{C}$, $(\mathrm{cog}_{D'}(G))(C) = \mathrm{colim}\ (G \circ D'(C))$ where $D'(C): \underline{A} \longrightarrow \underline{B} \times \underline{I}$ for some small category $\underline{A}$ (cf. (4.15)) while $C = \mathrm{colim}\ (D \circ D'(C))$. Thus to define a bijection

$$(6.26) \qquad \underline{E}((\mathrm{cog}_{D'}(G))(C), E) \cong \underline{C}(C, H(E))$$

it suffices to show that $\underline{E}\ (\mathrm{colim}\ (G \circ D'(C)), E) \cong \underline{C}(\mathrm{colim}\ (D \circ D'\ (C)), H(E))$, or equivalently, using a well known adjunction, to define one between $(\underline{A}, \underline{E})(G \circ D'(C), E_{\underline{A}})$ and $(\underline{A}, \underline{C})(D \circ D'(C), (H(E))_{\underline{A}})$. To this end, let $\alpha: G \circ D'(C) \Rightarrow E_{\underline{A}}$, i.e., for each X in $\underline{A}$, $\alpha(X): (G \circ D'(C))(X) \longrightarrow E$. Let $(D'(C))(X) = (B, I)$ and if $X \xrightarrow{x} X'$ in $\underline{A}$, let $(D'(C))(x) = (b, i)$. The naturality of α means that $\alpha(X') \cdot (G(D'(C))(x)) = \alpha(X)$. We then define $\alpha': D \circ D'(C) \Rightarrow (H(E))_{\underline{A}}$ by $\alpha'(X) = (\eta^{-1}(B, I, H(E)) \cdot \theta(B, I, E))(\alpha(X))$. To show that α' is natural, we must show that $\alpha'(X') \cdot (D(D'(C))(x)) = \alpha'(X)$ where again $X \xrightarrow{x} X'$ in $\underline{A}$, or equivalently, that $((\eta^{-1}(B', I', H(E)) \cdot \theta(B', I', E))\ \alpha(X')) \cdot D(b, i) = (\eta^{-1}(B, I, H(E)) \cdot \theta(B, I, E))(\alpha(X))$. But this follows from the commutativity of the following diagram

$$
\begin{array}{ccccc}
\underline{E}(G(B',I'),E) & \xrightarrow{\Theta(B',I',E)} & \underline{B}(B',G^{\#}(I',E)) & \xrightarrow{\eta^{-1}(B',I',H(E))} & \underline{C}(D(B',I'),H(E)) \\
\downarrow{\scriptstyle \underline{E}(G(b,i),1)} & & \downarrow{\scriptstyle \underline{B}(b,G^{\#}(i,1))} & & \downarrow{\scriptstyle \underline{C}(D(b,i),1)} \\
\underline{E}(G(B,I),E) & \xrightarrow{\Theta(B,I,E)} & \underline{B}(B,G^{\#}(I,E)) & \xrightarrow{\eta^{-1}(B,I,H(E))} & \underline{C}(D(B,I),H(E))
\end{array}
$$

and the correspondence $\alpha \rightsquigarrow \alpha'$ is a bijection since Θ and η^{-1} are. And this gives us the bijection (6.26). To show the latter is natural in C, let $C' \xrightarrow{c} C$, $D'(C'): A' \longrightarrow \underline{B} \rtimes \underline{I}$, and $D'(c) = (D_c, \bar{c})$, where $D_c: \underline{A}' \longrightarrow \underline{A}$ and $\bar{c}: D'(C') \Rightarrow (D'(C))\circ D_c$. Using the same detour as above (colim has a right adjoint), we must show that for any $\alpha: G\circ D_c \Rightarrow E_{\underline{A}'}$, $[(\alpha_* D_c)\cdot(G_* \bar{c})]' = (\alpha'_* D_c)\cdot(D_* \bar{c})$ where the " ' " keeps the same meaning as above. If X' is in $\underline{A}'$, we let $D_c(X') = X$, $(D'(C'))(X') = (B', I')$, $(D'(C))(X) = (B, I)$ and $\bar{c}(X') = (b', i'): (B', I') \longrightarrow (B, I)$. Now $((\alpha'_* D_c)(D_* \bar{c}))(X') = (\alpha'(X))\cdot D(b', b') = (\eta^{-1}(B, I, H(E)).$ $\Theta(B, I, E)\alpha(X))\cdot D(b', i')$, while $[(\alpha_* D_c)\cdot(G_* \bar{c})]'(X') = (\eta^{-1}(B', I', H(E))\cdot \Theta(B', I', E)\ (\alpha(X)\cdot G(b', i'))$, but this follows from the naturality of η and Θ (just reverse the vertical arrows in the above diagram, using b' and i' instead of b and i). If finally $E \xrightarrow{e} E'$, then $e\cdot\alpha'(X) = H(e)\cdot\alpha'(X)$, for all X in $\underline{A}$, from the naturality of η and Θ again, and thus we have naturality in the second variable. Q.E.D.

As mentioned above, the result also follows from Theorem 2.24 and remark 2.26 of [12] as follows: $D \dashv D^{\#}$ and $G \dashv D^{\#}\circ(\underline{I}_* \rtimes H)$

mean that for all E, I and B, $\underline{E}(G(B, I), D) \cong \underline{B}(B, D^{\#}(I, H(E))) \cong \underline{C}(D(B, I), H(E))$ natural in all variables, and this says precisely that G is D-left adjoint to H and thus the above mentioned theorem applies.

THEOREM 19. Keeping the same hypothesis as in the last theorem, let now K: $\underline{C} \longrightarrow \underline{E}$, H: $\underline{E} \longrightarrow \underline{C}$ and $\underline{E}$ be $K \circ D \circ D'(\underline{C})$-cocomplete. If K preserves the colimit of all diagrams of the form $D \circ D'(C)$ for all C in $\underline{C}$ and $K \circ D \dashv D^{\#} \circ (I_{\underline{I}^*} \times H)$, then $K \dashv H$ (and thus K preserves all colimits).

Proof. From Theorem 18, $K \circ D \dashv D^{\#} \circ (I_{\underline{I}^*} \times H)$ implies that $\mathrm{cog}_{D'}(K \circ D) \dashv H$. But Theorem 16 says that $\mathrm{cog}_{D'}(K \circ D) \cong K$. Q.E.D.

The duals of those two theorems are immediate.

THEOREM 20. Let D and K be as in theorem 19 while F: $\underline{J} \times \underline{X} \longrightarrow \underline{E}$ and H: $\underline{E} \longrightarrow \underline{C}$ satisfy the dual hypothesis. Then $K \dashv H$ iff for all I in $\underline{I}$ and J in $\underline{J}$, $F^{b}(\ , J) \circ K \circ D(\ , I) \dashv D^{\#}(I,\ ) \circ H \circ F(J,\ )$, where for all J, $F^{b}(\ , J) \dashv F(J,\ )$.

Proof. If the condition is satisfied and I is fixed, then by the dual of Theorem 18, $K \circ D(\ , I) \dashv \mathrm{glo}_{F'}(D^{\#}(I,\ ) \circ H \circ F)$, where the latter sends any E in $\underline{E}$ onto $\mathrm{Lim}(D^{\#}(I,\ ) \circ H \circ F \circ F'(E)) \cong D^{\#}(I,\ ) \circ ((\mathrm{glo}_{F'} \circ (F, I)')(H)) \cong D^{\#}(I,\ ) \circ H$ using the dual of Theorem 16 and the fact that $D^{\#}(I,\ )$ preserves limits. Now $K \circ D(\ , I) \dashv D^{\#}(I,\ ) \circ H$ for all I implies by Theorem 19 that K is left adjoint to H.

COROLLARY 1. Let D: $\underline{B} \longrightarrow \underline{C}$ be functorially codense and

D have a right adjoint $D^{\#}$. If H: $\underline{E} \longrightarrow \underline{C}$, G: $\underline{B} \longrightarrow \underline{E}$, $\underline{E}$ being $G \bullet D'(\underline{C})$-cocomplete and $G \dashv D^{\#} \bullet H$, then $\mathrm{cog}_{D'}(G) \dashv H$.

COROLLARY 2. With the same hypothesis on D as in the above corollary, let now K: $\underline{C} \longrightarrow \underline{E}$ satisfy that of Theorem 19. If $K \bullet D \dashv D^{\#} \bullet H$, then $K \dashv H$, and conversely.

COROLLARY 3. Let D: $\underline{B} \longrightarrow \underline{C}$ be as in Corollary 1 and dually for F: $\underline{X} \longrightarrow \underline{E}$, with $F^b \dashv F$. If H and K are as in Theorem 20 then $K \dashv H$ iff $F^b \bullet K \bullet D \dashv D^{\#} \bullet H \bullet F$.

For the next propositions, we recall the notation introduced after (1.1) and the convention introduced after the Corollary of Theorem 15.

PROPOSITION 21. Let H: $\underline{C} \longrightarrow (\underline{A}, \underline{B})$ where $\underline{B}$ is $|\underline{A}^2|$-copowered.

If $H_{\underline{A} \ltimes \underline{C}}$: $\underline{A} \ltimes \underline{C} \longrightarrow \underline{B}$ has a left adjoint G: $\underline{B} \rtimes \underline{A}^* \longrightarrow \underline{C}$ and $\underline{C}$ is $G \bullet (\overline{\Gamma^b})'(F)$-cocomplete for all F: $\underline{A} \longrightarrow \underline{B}$, (in particular if $\underline{C}$ is $\hat{\underline{A}}$-cocomplete) then H has a left adjoint equal to $K_{\Gamma^b}(G)$. Conversely, if H has a left adjoint H^b: $(\underline{A}, \underline{B}) \longrightarrow \underline{C}$, then $H_{\underline{A} \ltimes \underline{C}}$ has a left adjoint $H^b \bullet \Gamma^b$: $\underline{B} \rtimes \underline{A}^* \longrightarrow \underline{C}$ and $K_{\Gamma^b}(\overline{H^b \bullet \Gamma^b}) \cong H^b$. Thus H has a left adjoint iff $H_{\underline{A} \ltimes \underline{C}}$ has one.

<u>Proof</u>: Γ^b has a right adjoint Γ and the hypothesis implies that for all A in $\underline{A}$, $G(\ , A) \dashv H_{\underline{A} \ltimes \underline{C}}(A,\ ) = \Gamma_A \bullet H = \Gamma(A,\ ) \bullet H$. Thus the result follows from Theorem 18. Conversely, for all A in $\underline{A}$, $H^b \bullet \Gamma^b(\ , A) \dashv \Gamma(A,\ ) \bullet H = H_{\underline{A} \ltimes \underline{C}}(A,\ )$ and the result follows from Proposition 6 and the Corollary of Theorem 16. Q.E.D.

This would also follow directly from Theorem 2.24 of [12] since Γ^b is dense in the sense of Ulmer by Corollary 3 of Proposition 9 and Proposition 10 (without any "smallness" condition on $\underline{B}$).

Now the following version of the preceding Proposition can be proved directly without using Theorem 18.

PROPOSITION 22. If H: $\underline{C} \to (\underline{A}, \underline{B})$ where $\underline{C}$ is $\hat{\underline{A}}$-cocomplete and $H_{\underline{A}\ltimes\underline{C}}$: $\underline{A}\ltimes\underline{C} \to \underline{B}$ has a left adjoint G: $\underline{B}\ltimes\underline{A}^* \to \underline{C}$ then H has a left adjoint H^b where for any F: $\underline{A} \to \underline{B}$, $H^b(F)$ is the colimit of $\hat{\underline{A}} \xrightarrow{(\Gamma^b)'(F)} \underline{B}\ltimes\underline{A}^* \xrightarrow{G} \underline{C}$.

<u>Proof</u>. The proof is as in Theorem 18 but simpler. We are given $\Theta(B, A, C)$: $\underline{C}(G(B, A), C) \cong \underline{B}(B, (H(C))(A))$ and to show that $\underline{C}(\text{colim } (G\circ(\Gamma^b)'(F), C) \cong (\underline{A}, \underline{B})(F, H(C))$ it suffices to prove that $(\hat{\underline{A}}, \underline{C})(G\circ(\Gamma^b)'(F), C_{\hat{\underline{A}}}) \cong (\underline{A}, \underline{B})(F, H(C))$: if ϕ: $G\circ(\Gamma^b)'(F) \to C_{\hat{\underline{A}}}$, define $\overline{\phi}$: f $\to$ H(C) by $\overline{\phi}(A) = \Theta(F(A), A, C)$. The rest is straightforward. Q.E.D.

But we hasten to point out that H^b is not in that case the Γ^b-Kan extension of G since Γ^b: $\underline{B}\ltimes\underline{A}^* \to (\underline{A}, \underline{B})$ does not necessarily exist, $\underline{B}$ not being $|\underline{A}^2|$-copowered.

PROPOSITION 23. Let $\underline{B}$ be $|\underline{A}^2|$-copowered, K: $(\underline{A}, \underline{B}) \to \underline{C}$ preserve the colimit of all diagrams of the form $\Gamma^b\circ(\Gamma^b)'(F)$ and $\underline{C}$ be $K\circ\Gamma^b\circ(\Gamma^b)'(F)$-cocomplete for all F: $\underline{A} \to \underline{B}$ (in particular when $\underline{C}$ is $\hat{\underline{A}}$-cocomplete and K preserves $\hat{\underline{A}}$-colimits). If $K\circ\Gamma^b$ has right adjoint H: $\underline{A}\ltimes\underline{C} \to \underline{B}$, then K is left adjoint to $H_{\underline{C}}$: $\underline{C} \to (\underline{A}, \underline{B})$. Conversely, if K has a right adjoint $K^{\#}$, then $K\circ\Gamma^b \dashv K_{\underline{A}\times\underline{C}}$ $(K_{\Gamma^b}(K\circ\Gamma^b) \cong K$ and K preserves all colimits. Thus K has a right adjoint iff it preserves colimits and $K\circ\Gamma^b$ has a right adjoint.

<u>Proof</u>. The hypothesis implies that for all A, $K\circ\Gamma^b(\ ,A) \dashv H(A,\ ) = \Gamma(\ ,A)\circ H_{\underline{C}}$ and thus Theorem 19 applies. The converse is obvious, using Proposition 6 and the Corollary of Theorem 16. Q.E.D.

The last two propositions can be combined to give the next one which is for simplicity stated in the "$\hat{\underline{A}}$-terminology" as will also be the six results following it:

PROPOSITION 24. If $\underline{A}$ and $\underline{C}$ are small, if $\underline{B}$ is $|\underline{A}^2|$-copowered and is $(\hat{\underline{A}})^*$-complete and dually for $\underline{D}$ (w.r. to $\underline{C}$), if K: $(\underline{A}, B) \to (\underline{C}, \underline{D})$ preserves $\hat{\underline{A}}$-colimits and if $(K\circ\Gamma^b)_{\underline{B}\times(\underline{A}^*\times\underline{C})}$ has a right adjoint H: $(\underline{A}\times\underline{C}^*)\times\underline{D} \to \underline{B}$, then K has a right adjoint equal to $K_{\Gamma^\#}(H_{\underline{C}^*\times\underline{D}})$. Conversely, if K has a right adjoint, then K preserves colimits and $(K\circ\Gamma^b)_{\underline{B}\times(\underline{A}^*\times\underline{C})} \dashv (K^\#\circ\Gamma^\#)_{(\underline{A}\times\underline{C}^*)\times\underline{D}}$. Thus if G: $(\underline{C}, \underline{D}) \to (\underline{A}, \underline{B})$, then $K \dashv G$ iff for all A and C, $\Gamma_C\circ K\circ\Gamma^b_A \dashv \Gamma_A\circ G\circ\Gamma^\#_C$, and K preserves colimits and dually for G.

<u>Proof</u>. The hypothesis implies that for a fixed C in $\underline{C}$, $\Gamma_C\circ K\circ\Gamma^b \dashv H(\ ,C,\ ): \underline{A}\times\underline{D} \to \underline{B}$. Thus Proposition 23 applies to $\Gamma_C\circ K$: $(\underline{A}, \underline{B}) \to \underline{D}$, and it has a right adjoint $(H(\ ,C,\ ))_{\underline{D}}$: $\underline{D} \to (\underline{A}, \underline{B})$. By Proposition 6, $K_{(\underline{A},\underline{B})\times\underline{C}}$: $(\underline{A}, \underline{B})\times\underline{C} \to \underline{D}$ has a right adjoint $H_{\underline{C}^*\times\underline{D}}$: $\underline{C}^*\times\underline{D} \to (A, \underline{B})$. Using the dual of Proposition 21, K is left adjoint to $K_{\Gamma^\#}(H_{\underline{C}^*\times\underline{D}})$: $(\underline{C}, \underline{D}) \to (\underline{A}, \underline{B})$.

LEMMA. If $G: \underline{B} \times \underline{A}^* \longrightarrow \underline{C}$ preserves colimits, $\underline{B}$ is cocomplete and $\underline{C}$ is $\hat{\underline{A}}$-cocomplete, then $K_{\Gamma^b}(G)$ preserves colimits.

Proof. $K_{\Gamma^b}(G)$ is equal to the following composition: $(\underline{A}, \underline{B}) \xrightarrow{(\Gamma^b)'} (\hat{\underline{A}}, \underline{B} \times \underline{A}^*) \xrightarrow{(I, G)} (\hat{\underline{A}}, \underline{C}) \xrightarrow{\text{colim}} \underline{C}$ and it is easy to see that $(\Gamma^b)'$ commutes with all colimits, just as (I, G) and of course colim.

PROPOSITION 25. If $\underline{B}$ is cocomplete and $\underline{C}$ is $\hat{\underline{A}}$-cocomplete, then the following (meta)-functor:

$$((\underline{A}, \underline{B}), \underline{C})_{c\ell} \xrightarrow{(\Gamma^b, I)} (\underline{B} \times \underline{A}^*, \underline{C})_{c\ell} \xrightarrow{K_{\Gamma^b}} ((\underline{A}, \underline{B}), \underline{C})_{c\ell}$$

where the index "$c\ell$" denotes the full (meta)-subcategories of all functors preserving colimits, is naturally equivalent to the identity.

Proof. Follows from above lemma and Theorem 16. Q.E.D.

The proof of the next proposition is adapted from [1]:

PROPOSITION 26. With same restrictions on $\underline{B}$ and $\underline{C}$ as above and the same notation, the (meta)-functor:

$$(\underline{B} \times \underline{A}^*, \underline{C})_{c\ell} \xrightarrow{K_{\Gamma^b}} ((\underline{A}, \underline{B}), \underline{C})_{c\ell} \xrightarrow{(\Gamma^b, I)} (\underline{B} \times \underline{A}^*, \underline{C})_{c\ell}$$

is naturally equivalent to the identity.

Proof. We first show that the functor:

$$(\underline{A}^*, \underline{C}) \times \underline{A}^* \xrightarrow{\Gamma^b} (\underline{A}, (\underline{A}^*, \underline{C})) \xrightarrow{(\Gamma^b)'} (\hat{\underline{A}}, ((\underline{A}^*, \underline{C}) \times \underline{A}^*)) \xrightarrow{(I, \Gamma)} (\hat{\underline{A}}, \underline{C}) \xrightarrow{\text{colim}} \underline{C}$$

is naturally equivalent to Γ. Let D: $\underline{A}^* \longrightarrow \underline{C}$ and A" be in $\underline{A}$. Then the above composition applied to (D, A") is equal to the colimit of the functor

$$\hat{\underline{A}} \longrightarrow (\underline{A}^*,\underline{C})\rtimes\underline{A}^* \xrightarrow{\Gamma} \underline{C}\colon \hat{a} \rightsquigarrow (\underline{A}(A'',A)\cdot D, A') \rightsquigarrow \underline{A}(A'',A)\cdot D(A')$$

where $A \xrightarrow{a} A'$ in $\underline{A}$, and this is precisely D(A") = Γ(D, A") using Theorem 9 and the fact that $\Gamma_{A''}$ preserves colimits.

Let now G: $\underline{B}\rtimes\underline{A}^* \longrightarrow \underline{C}$. Then $G_{\underline{B}}\colon \underline{B} \longrightarrow (\underline{A}^*, \underline{C})$ and we obtain

$$\begin{array}{ccccccccc}
(\underline{A}^*,\underline{C})\rtimes\underline{A}^* & \xrightarrow{\Gamma^b} & (\underline{A},(\underline{A}^*,\underline{C})) & \xrightarrow{(\Gamma^b)'} & (\hat{\underline{A}},((\underline{A}^*,\underline{C})\rtimes\underline{A}^*)) & \xrightarrow{(I,\Gamma)} & (\underline{A},\underline{C}) & \xrightarrow{\text{colim}} & \underline{C} \\
\uparrow{\scriptstyle G_{\underline{B}}\rtimes I} & & \uparrow{\scriptstyle (1,G_{\underline{B}})} & & \uparrow{\scriptstyle (1,G_{\underline{B}}\rtimes I)} & \nearrow{\scriptstyle (I,G)} & & & \\
\underline{B}\rtimes\underline{A}^* & \xrightarrow{\Gamma^b} & (\underline{A},\underline{B}) & \xrightarrow{(\Gamma^b)'} & (\hat{\underline{A}}, \underline{B}\rtimes\underline{A}^*) & & & &
\end{array}$$

where the squares and the triangle commute. Thus, the top line composed with $G_{\underline{B}}\rtimes I$ is equal to the bottom path, and the former is precisely G using the first part of the proof. Similarly for the morphisms.

COROLLARY 1. If $\underline{B}$ is cocomplete and $\underline{C}$ is $\hat{\underline{A}}$-cocomplete, then $(\Gamma^b, I)\colon ((\underline{A}, \underline{B}), \underline{C})_{c\ell} \longrightarrow ((\underline{B}\rtimes\underline{A}^*), \underline{C})_{c\ell}$ is a (meta)-equivalence, and similarly for K_{Γ^b}.

COROLLARY 2. With the same conditions on $\underline{B}$ and $\underline{C}$, there is a (meta)-equivalence $((\underline{A}, \underline{B}), \underline{C})_{c\ell} \simeq (\underline{B}, (\underline{A}^*, \underline{C}))_{c\ell}$, "natural in all variables", and given by the isomorphism (1.1) composed with (Γ^b, I).

In fact this looks like a relative (meta)-adjointness relation "up to a natural equivalence" for ($\underline{A}$,) and ($\underline{A}^*$,)!

COROLLARY 3. Replacing in the above corollary "$c\ell$" by "rad" for the full subcategories of functors having a right adjoint, gives another equivalence

$$((\underline{A}, \underline{B}), \underline{C})_{rad} \simeq (\underline{B}, (\underline{A}^*, \underline{C}))_{rad}$$

"natural in all variables" where $\underline{C}$ is $\hat{\underline{A}}$-cocomplete and is $|\underline{A}^2|$-powered, and dually for $\underline{B}$.

Proof. That (Γ^b, I) "restricts" to the full subcategory on the left follows from Proposition 22 and it is an equivalence from the above Corollary 1. The dual of Proposition 21 says precisely that the isomorphism (1.1) "restricts" to the full subcategories of functors having a right adjoint. Q.E.D.

We end this section with another application of Proposition 21, but before, we need the following:

PROPOSITION 27. If $T: \underline{A} \longrightarrow \underline{B}$, then $K_Y(T) = K_{\Gamma^b}(\sum\bullet(I \times T))$: $(\underline{A}^*, \text{Ens}) \longrightarrow \underline{B}$, where $\sum$: $\text{Ens} \times \underline{B} \longrightarrow \underline{B}$ and Γ^b: $\text{Ens} \times \underline{A} \longrightarrow (\underline{A}^*, \text{Ens})$.

Proof. For any $D: \underline{A}^* \longrightarrow \text{Ens}$ and $A \xrightarrow{a} A'$, by definition, $(K_Y(T))(D)$ is the colimit of the functor $\hat{\underline{A}} \longrightarrow \underline{B}$:
$\hat{a} \rightsquigarrow (T(A), A') \rightsquigarrow (T(A), \underline{A}(\ ,A)) \rightsquigarrow \text{Nat}(\underline{A}(\ ,A'), D)\cdot(T(A)) \simeq (D(A'))\cdot(T(A))$, using the Yoneda Lemma, and this is precisely $(K_{\Gamma^b}(\sum\bullet(1 \times T)))(D)$. Same for the morphisms. Q.E.D.

In fact, if $\sum$: $\text{Ens} \times (\underline{A}, \text{Ens}) \longrightarrow (\underline{A}, \text{Ens})$, then Theorem 9 and Proposition 8 say that $K_{\Gamma^b}(\sum\bullet(1 \times Y^*))$: $(\underline{A}, \text{Ens}) \longrightarrow (\underline{A}, \text{Ens})$

is naturally equivalent to the identity endofunctor. But the above result says that $K_{Y^*}(Y^*) = K_{\Gamma^b}(\sum\circ(1\times Y^*))$, and this gives Proposition 11 (for $\underline{A}^*$).

PROPOSITION 28. Let $T: \underline{A} \longrightarrow \underline{B}$, where $\underline{B}$ has all coproducts of the form $E\cdot T(A)$ for all sets E and objects A in $\underline{A}$ and also all colimits of diagrams of the form $\sum\circ(1\times T)\circ(\Gamma^b)'(D)$ for all $D: \underline{A}^* \longrightarrow \mathrm{Ens}$. Then $\overline{T}: \underline{B} \longrightarrow (\underline{A}^*, \mathrm{Ens})$ (c.f. 1.4) has a left adjoint equal to $K_{\Gamma^b}(\sum\circ(1\times T)) = K_Y(T)$, where $(K_Y(T))(D)$ is the colimit of the functor

$(\underline{A}^*, \mathrm{Ens})_r/D \xrightarrow{Y'(D)} \underline{A} \xrightarrow{T} \underline{B}$: $\underline{A}(A, \ ) \longrightarrow D \rightsquigarrow T(A)$ (see 4.19), for all $D: \underline{A}^* \longrightarrow \mathrm{Ens}$.

Proof. $\mathrm{Ens}\times\underline{A} \xrightarrow{1\times T} \mathrm{Ens}\times\underline{B} \xrightarrow{\sum} \underline{B}$ (in fact one can go directly from $\mathrm{Ens}\times\underline{A}$ to $\underline{B}$ but we keep the above notation for simplicity), is the left adjoint of $\overline{T}_{\underline{A}^*\times\underline{B}}: \underline{A}^*\times\underline{B} \longrightarrow \mathrm{Ens}$ since we have $\underline{B}(E\cdot(T(A)), B) \cong \underline{B}(T(A), B)^E \cong \mathrm{Ens}(E, \underline{B}(T(A), B))$ and $\overline{T}_{\underline{A}^*\times B}(A, B) = \underline{B}(T(A), B)$. The result follows from Propositions 14, 21 and 27. Q.E.D.

From Proposition 27, it is sufficient that $\underline{B}$ contain the colimit of all diagrams of the form $T\circ Y'(D)$ for all D.

COROLLARY 1. If $\underline{A}$ has all copowers and is $\sum\circ(\Gamma^b)'(D)$-cocomplete for all $D: \underline{A}^* \longrightarrow \mathrm{Ens}$, and $\sum: \mathrm{Ens}\times\underline{A} \longrightarrow \underline{A}$, then the Yoneda functor $Y: \underline{A} \longrightarrow (\underline{A}^*, \mathrm{Ens})$ has a left adjoint equal to $K_{\Gamma^b}(\sum) = K_Y(1_{\underline{A}})$, which means there is a bijection

$\underline{A}((K_Y(I_{\underline{A}}))(D), A) \cong (\underline{A}^*, Ens)(D, \underline{A}(\;, A))$ natural in $D: \underline{A}^* \longrightarrow Ens$ and A in $\underline{A}$. (See the Yoneda Lemma!)

COROLLARY 2. If $\underline{A}$ and $\underline{B}$ are small, $\underline{B}$ has all copowers and is $\sum\circ(1\times\Gamma^b)\circ(\Gamma_o^b)'(D)$-cocomplete for all $D: (B\times A^*)^* \longrightarrow Ens$ where Γ_o^b: $Ens\times(\underline{B}\times\underline{A}^*) \longrightarrow ((\underline{B}\times\underline{A}^*)^*, Ens)$ and Y_o: $(\underline{B}\times\underline{A}^*) \longrightarrow ((\underline{B}\times\underline{A}^*)^*, Ens)$, then the embedding Γ^b: $(\underline{A}, \underline{B}) \longrightarrow ((B\times A^*)^*, Ens)$ (c.f. Theorem 13) has a left adjoint equal to $K_{\Gamma_o{}^b}(\sum\circ(1\times\Gamma^b)) = K_{Y_o}(\Gamma^b)$, where $Ens\times\underline{B}\times\underline{A}^* \xrightarrow{1\times\Gamma^b} Ens\times(\underline{A}, \underline{B}) \xrightarrow{\sum} (\underline{A}, \underline{B})$.

7. The Kan Extension Theorems.

In this section, we show how the various forms of the Kan Extension Theorem can be obtained from Proposition 21 and finally prove the promised equivalence in a series of Remarks.

PROPOSITION 29. If $\underline{A}$ is small, $\underline{B}$ has all copowers of the form $E\cdot(F(\underline{A}(\;, A)))$ for all sets E and objects A in $\underline{A}$, and is $F\circ\Gamma^b\circ(\Gamma^b)'(D)$-cocomplete for all $D: \underline{A}^* \longrightarrow Ens$, where $F: (\underline{A}^*, Ens) \longrightarrow \underline{B}$, then F has a right adjoint equal to $\overline{F\circ Y}$ iff F commutes with colimits. If $\underline{B}$ is cocomplete then $((\underline{A}^*, Ens), \underline{B})_{c\ell} \underset{K_Y}{\overset{(Y, I)}{\rightleftarrows}} (\underline{A}, \underline{B})$ is a (meta)-equivalence of categories.

Proof. $F\circ\Gamma^b$: $Ens\times\underline{A} \longrightarrow \underline{B}$: $(E, A) \rightsquigarrow \underline{A}(\;, A)\cdot E \cong E\cdot A(\;, A) \rightsquigarrow E\cdot F(\underline{A}(\;, A))$ is the left adjoint of $(\overline{F\circ Y})_{\underline{A}^*\times\underline{B}}$ and the first result follows from Proposition 23. If

$T: \underline{A} \longrightarrow \underline{B}$, then $K_Y(T) \dashv \overline{T}$ by Proposition 28, thus $K_Y(T)$ preserves colimits and the equivalence follows from the Corollary of Theorem 17. Q.È.D.

In fact, the above theorem says that $((\underline{A}^*, \text{Ens}), \underline{B})_{c\ell} = ((\underline{A}^*, \text{Ens}), \underline{B})_{rad}$ (when $\underline{B}$ is cocomplete, say) and by setting $\underline{A} = \underline{1}$, that a functor $G: \underline{B} \longrightarrow \text{Ens}$ has a left adjoint iff it is representable by some B in $\underline{B}$ where $\underline{B}$ must have all copowers $E \cdot B$ for all sets E. For the next Proposition we recall Part 1 of the Corollary of Proposition 14.

PROPOSITION 30. If $F: \underline{A} \longrightarrow \underline{B}$, both categories small, $\underline{C}$ is $|\underline{A}^2|$-copowered and satisfies (5.24) and (5.23) for all $H: \underline{A} \longrightarrow \underline{C}$, then $(F, I): (\underline{B}, \underline{C}) \longrightarrow (\underline{A}, \underline{C}): D \rightsquigarrow D \circ F$ is the right adjoint of K_F and $K_F = K_{\Gamma^b}(\Gamma^b \circ (1 \ltimes F^*_*))$.

<u>Proof</u>. This follows from the fact that $(F, I)_{\underline{A} \ltimes (\underline{B}, \underline{C})}$: $\underline{A} \ltimes (\underline{B}, \underline{C}) \longrightarrow C: (A, D) \rightsquigarrow D(F(A) = \Gamma_{F(A)}(D)$ is the right adjoint of $\underline{C} \ltimes \underline{A}^* \xrightarrow{1 \ltimes F^*_*} \underline{C} \ltimes \underline{B} \xrightarrow{\Gamma^b} (\underline{B}, \underline{C})$, from Proposition 21 and Corollary (1) of Proposition 14. Q.E.D.

The above proof is essentially due to Andre ([1]). Now from Proposition 22 the following version of the preceding Proposition can be obtained:

PROPOSITION 31. If $\underline{A}$ and $\underline{B}$ are small, $\underline{C}$ is $\hat{\underline{A}}$-cocomplete and satisfies (5.22) for all $H: \underline{A} \longrightarrow \underline{C}$, then (F, I) is the right adjoint of K_F.

<u>Proof</u>. Same as above. Q.E.D.

But the remark made after Proposition 22 also applies here: K_F will not in general be equal to $K_{\Gamma^b}(\Gamma^b \circ (1 \times F^*_*))$, (see again the convention made in Corollary (1) of Proposition 14), for Γ^b: $\underline{C} \times \underline{A}^* \longrightarrow (\underline{A}, \underline{C})$ need not exist since $\underline{C}$ is not $|\underline{A}^2|$-copowered.

From Proposition 30 comes the "abstract" definition of a Kan extension: $K_F \dashv (F, I)$ means that for any H: $\underline{A} \longrightarrow \underline{C}$ and D: $\underline{B} \longrightarrow \underline{C}$ there is a bijection $(\underline{B}, \underline{C})(K_F(H), D) \cong (\underline{A}, \underline{C})(H, (F, I)(D))$ natural in H and D.

In the general situation $\underline{B} \xleftarrow{F} \underline{A} \xrightarrow{H} \underline{C}$, the (right) Kan F-extension of H is any functor $K_F(H)$: $\underline{B} \longrightarrow \underline{C}$ such that for any other functor D: $\underline{B} \longrightarrow \underline{C}$ there exists a bijection $\mathrm{Nat}(K_F(A), D) \cong \mathrm{Nat}(H, D \circ F)$ natural in H and D, Nat standing for the class of natural transformations.

From the properties of adjointness $K_F(H)$ is unique up to natural equivalence when it exists and Proposition 30 gives sufficient conditions for the existence of the "abstract" Kan F-extensions.

As noted before $\underline{B}$ need not be small and if large diagrams are allowed neither does $\underline{A}$. (5.24) can also be replaced by (5.22) but again K_F is no longer equal to $K_{\Gamma^b}(\Gamma^b \circ (1 \times F^*_*))$ for the functor (essentially $\Gamma^b \circ (1 \times F^*_*)$ when $\underline{C}$ is $|\underline{A}^2|$-copowered) $\underline{C} \times \underline{A}^* \longrightarrow (\underline{B}, \underline{C})$: $(C, A) \leadsto \underline{B}(F(A), \ ) \cdot C$ need not exist. In fact it is claimed by Ulmer ([12]) that $\underline{C}$ need not be $|\underline{A}^2|$-copowered: this is necessary here if Γ^b: $\underline{C} \times \underline{A}^* \longrightarrow (\underline{A}, \underline{C})$ is to exist and Γ^b is needed for in the definition of functorial codensity it is required that $\hat{\underline{A}} \longrightarrow (\underline{A}, \underline{C}): \hat{a} \rightsquigarrow \underline{A}(A',)\cdot H(A)$, where $A \xrightarrow{a} A'$, factorize through Γ^b. He simply assumes that $\underline{C}$ has all coproducts of the form $\underline{A}(A', A")\cdot H(A)$ for all A, A' and A" in $\underline{A}$. Ulmer also claims that assuming that $\underline{C}$ satisfies (5.22) and the latter condition then (5.25) holds when the Kan F-extension of H exists.

The dual case is obvious and we now introduce the notion of "lifted functor" due to Kan ([5]).

Let $F: \underline{B} \times \underline{C} \longrightarrow \underline{D}$ and $\underline{A}$ be a small category. Then the functor $F_{\underline{B}}: \underline{B} \longrightarrow (\underline{C}, \underline{D})$ obtained through (1.1) gives $(\underline{A}, \underline{B}) \xrightarrow{(\overline{I}, F_{\underline{B}})} (\underline{A}, (\underline{C}, \underline{D})) \cong (\underline{C}, (\underline{A}, \underline{D}))$ which in turn gives by (1.1) again a functor ${}^{\underline{A}}F: (\underline{A}, \underline{B}) \times \underline{C} \longrightarrow (\underline{A}, \underline{D}): (H, C) \rightsquigarrow F\circ(H \times I)$, and F is said to be lifted in the first variable. For instance, ${}^{\underline{A}}{}^{*}(\underline{B}(,))(H,) = \overline{H}$, if $H: \underline{A} \longrightarrow \underline{B}$. ($\overline{H}$ defined in 1.4).

One could also lift in the second variable to obtain $F^{\underline{A}}: \underline{B} \times (\underline{A}, \underline{C}) \longrightarrow (\underline{A}, \underline{D}): (B, G) \rightsquigarrow F\circ(1 \times G)$. If $\underline{A}$ and $\underline{A}'$ are both small, one obtains a functor ${}^{\underline{A}'}F^{\underline{A}}$ equal to the following composition:

$$(\underline{A}', \underline{B}) \times (\underline{A}, \underline{C}) \xrightarrow{({}^{\underline{A}'}F)^{\underline{A}}} (\underline{A}, (\underline{A}', \underline{D})) \cong (\underline{A}' \times \underline{A}, \underline{D}): (H, G) \rightsquigarrow F\circ(H \times G)$$

or equivalently to:

$$(\underline{A}', \underline{B}) \times (\underline{A}, \underline{C}) \xrightarrow{{}^{\underline{A}'}(F^{\underline{A}})} (\underline{A}', (\underline{A}, \underline{D})) \cong (\underline{A}' \times \underline{A}, \underline{D}).$$

PROPOSITION 32. Let $F: \underline{B}\times\underline{C} \longrightarrow \underline{D}$ have a right adjoint $F^{\#}: \underline{C}^*\times\underline{D} \longrightarrow \underline{B}$, $\underline{A}$ be a small category, $\underline{B}$ be $|\underline{A}^2|$-copowered and $\underline{D}$ be $F\circ(I\times G)\circ(\Gamma^b)'(H)$-cocomplete for all $H: \underline{A}^* \longrightarrow \underline{B}$. Then $^{\underline{A}^*}F: (\underline{A}, \underline{C})^*\times\underline{D} \longrightarrow (\underline{A}^*, \underline{B})$ has a left adjoint: $(\underline{A}^*, \underline{B})\times(\underline{A}, \underline{C}) \longrightarrow \underline{D}$: $(H, G) \rightsquigarrow (K_{\Gamma^b}(F\circ(1\times G)))(H)$ (obtained from $K_{\Gamma^b}(F\circ(1\times G))$ by Proposition 6) which is also equal to the following composition:

$$(\underline{A}^*, \underline{B})\times(\underline{A}, \underline{C}) \xrightarrow{\;^{\underline{A}^*}F^{\underline{A}}\;} (\underline{A}\times\underline{A}^*, \underline{D}) \xrightarrow{(K_o\circ T_o,\, 1)} (\hat{\underline{A}}, \underline{D}) \xrightarrow{\text{colim}} \underline{D}$$

<u>Proof</u>. $F \dashv F^{\#}$ implies that for all A and $G: \underline{A} \longrightarrow \underline{C}$, $F(\ , G(A)) \dashv F^{\#}(G(A), \ )$, and by Proposition 6, that $F\circ(1\times G)$: $\underline{B}\times\underline{A} \longrightarrow \underline{D}$ is the left adjoint of $F^{\#}\circ(G^*_*\times I): \underline{A}^*\times\underline{D} \longrightarrow \underline{B}$. By Proposition 21, $(F^{\#}\circ(F^*_*\times 1))_{\underline{D}} = (^{\underline{A}^*}F^{\#})(G, \ ): \underline{D} \longrightarrow (\underline{A}^*, \underline{B})$ has as left adjoint $K_{\Gamma^b}(F\circ(I\times G))$. Q.E.D.

Again the next Proposition is proved as above from Proposition 22:

PROPOSITION 33. If $F: \underline{B}\times\underline{C} \longrightarrow \underline{D}$ has a right adjoint $F^{\#}: \underline{C}^*\times\underline{D} \longrightarrow \underline{B}$, $\underline{A}$ is small and $\underline{D}$ is $\hat{\underline{A}}$-cocomplete, then $^{\underline{A}^*}F^{\#}$ has a left adjoint equal to the above composition.

The dual version is again immediate. All the theorems encountered so far in this section are usually referred to as Kan Extension Theorems. Proposition 31 is Lawvere's version ([7]) except that $\underline{C}$ is assumed to be cocomplete and Proposition 33 is given in ([5]).

We now wish to prove the equivalence announced in the

introduction. In fact we will only sketch the proof and make it a series of remarks since proving a real equivalence would involve setting different axiom systems for the "category of categories" and showing that the various axiom systems are equivalent.

<u>REMARK 1</u>. Theorems 7 and 9, Propositions 21, 23, 29, 30, 32 and the Corollaries of Theorems 16 and 17 are all (at least roughly speaking!) equivalent.

<u>Proof</u>. We will refer to the various propositions above by their numbers only.

$7 \Rightarrow 9$ was proved in 9, and 16, can easily be established from 9 without using Theorem 16. Conversely $16 \Rightarrow 9$ since letting $I: (\underline{A}, \underline{B}) \longrightarrow (\underline{A}, \underline{B})$ be the identity functor then $K_{\Gamma^b}(\Gamma^b) \cong I$ by 16, and thus for any $F: \underline{A} \longrightarrow \underline{B}$, $(K_{\Gamma^b}(\Gamma^b))(F) \cong F$ which is precisely 9. Now 9 and 16 combined imply 21, as shown in 21 while $21 \Rightarrow 30$ was proved in 30. $21 \Rightarrow 32$ as proved in 32 and the converse is also true for: if $H_{\underline{A} \times \underline{C}}: \underline{A} \times \underline{C} \longrightarrow \underline{B}$ has a left adjoint $G: \underline{B} \times \underline{A}^* \longrightarrow \underline{C}$, then $H = {}^{\underline{A}}(H_{\underline{A} \times \underline{C}})(1_{\underline{A}},\ ): \underline{C} \longrightarrow (\underline{A}, \underline{B})$ has a left adjoint which is precisely $K_{\Gamma^b}(G)$. $30 \Rightarrow 7$ since for each A in $\underline{A}$, $\epsilon_A: \underline{1} \longrightarrow \underline{A}: 1 \leadsto A$ induces $\Gamma_A = (\epsilon_A, I): (\underline{A}, \underline{B}) \longrightarrow (\underline{1}, \underline{B}) \cong \underline{B}$ where $\underline{B}$ is of course $|\underline{1}^2|$-copowered and trivially satisfies (5.24) while (5.23) holds since $\underline{B}$ is assumed to be $|\underline{A}^2|$-copowered. Now 21 and 16 both imply 23 as shown in 23 (but $21 \Leftrightarrow 16$ was proved above) and both 23 and 16 again imply 29, thus 9 implies 29. The converse is also true: from the proof of 29 one obtains 17 (in fact 29 could be restated in the language of 17) and the latter says that for any $F: \underline{A} \longrightarrow \underline{B}$, $K_Y(F) \bullet Y = K_{\Gamma^b}(\sum^{\bullet}(1 \times F_*^*)) \bullet Y \cong F$ (using of course Proposition 27 which is really a lemma) and this is precisely 9.

REMARK 2. If $\underline{B}$ is also small ($\underline{A}$ always is) then the following three groups of propositions are equivalent:

(1) Theorem 9

(2) Theorems 12 and 13

(3) Theorem 12 and: $\overline{\Gamma}^b$ (c.f. (4.21)) is a full embedding.

Proof. (1) $\Longleftrightarrow$ (2): Theorem 7 implies the Yoneda Lemma and both imply Theorem 13 as shown in the proof of that Theorem, while Proposition 12 was shown to follow from Proposition 11, the latter being proved from Theorem 9, which implies Theorem 7. The converse follows by applying Proposition 12 to Γ^b in order to obtain Theorem 9 (or at least its Corollary 3). Q.E.D.

(2) $\Longleftrightarrow$ (3): $\overline{\Gamma}^b$ is a full embedding iff Γ^b is left adequate.

N.B. If "left adequate" is replaced by "dense" in the sense of Ulmer then $\underline{B}$ above need not be small.

That the Kan Extension Theorems are all equivalent has been part of the folklore for quite a while and that the latter are equivalent to the "André Theorems" has been (orally) conjectured by Bénabou.

8. The Results of André.

In this final section, we show the relationship between the results of André ([1]) and those of sections 5, 6 and 7.

The following definition is from [1]:

DEFINITION 8. If $\underline{A}$ is small and in

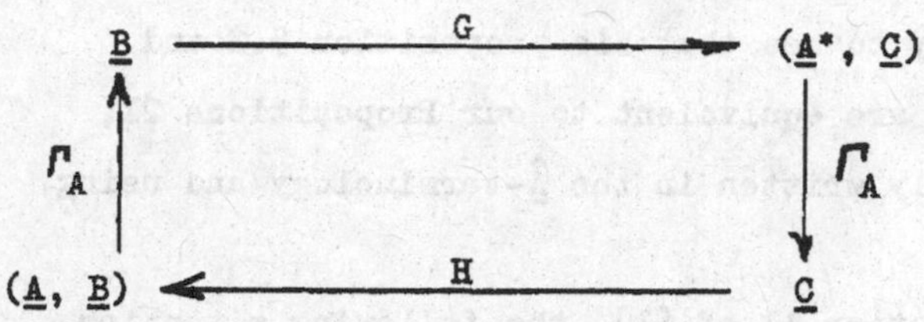

$\Gamma_A \bullet G$ is the left (right) adjoint of $\Gamma_A \bullet H$ for all A in $\underline{A}$, then G is said to be locally left (right) adjoint to H.

LEMMA. Keeping the same notation as above, G is locally left (right) adjoint to H iff $G_{\underline{A}^* \times \underline{B}}: \underline{A}^* \times \underline{B} \to \underline{C}$ is the left (right) adjoint of $H_{\underline{C} \times \underline{A}}: \underline{C} \times \underline{A} \to \underline{B}$ (using (1.1)).

<u>Proof</u>. Obvious.

This fact was communicated to me by Bénabou, and it is now easy to translate André's results into the terminology of the present paper:

$\Gamma^b: \underline{B} \times \underline{A}^* \to (\underline{A}, \underline{B})$ yields through (1.1) the functor $^*J = \Gamma^b_{\underline{B}}: \underline{B} \to (\underline{A}^* \times \underline{A}, \underline{B})$ of [1]. If F: $(\underline{A}, \underline{B}) \to \underline{C}$, the right localisation of F is the functor $(F \bullet \Gamma^b)_{\underline{B}}: \underline{B} \to (\underline{A}^*, \underline{C})$ obtained from $F \bullet \Gamma^b: \underline{B} \times \underline{A}^* \to \underline{C}$, while if G: $\underline{B} \to (\underline{A}^*, \underline{C})$, the right globalisation of G is equal to $K_{\Gamma^b}(G_{\underline{B} \times \underline{A}^*})$, and the fact that globalisation and localisation are "inverse constructions" is expressed in our Propositions 25 and 26. Thus "localizing" the identity endofunctor of $(\underline{A}, \underline{B})$ and then "globalising" it gives precisely Theorem 9 as shown in 16 $\Rightarrow$ 9 of Remark 1 or its Corollary 3 in the terminology of André, since he uses $(\underline{\tilde{A}})^*$ (in fact $\tilde{A}$ in [1], but this must be a misprint) instead of $\underline{\hat{A}}$. It is now easy to see that his proposition 5.2 and theorems 6.1 and 7.1 are equivalent to our Propositions 21, 23 and 24 respectively written in the $\underline{\hat{A}}$-terminology and using his remark 6.3.

Finally in section 11 of [1], the following situation is studied: let $H: \underline{B}^* \times \underline{A} \longrightarrow Ens$ (a profunctor from $\underline{B}$ to $\underline{A}$), $\underline{A}$ and $\underline{B}$ small, $\underline{C}$ complete and cocomplete. Then $\sum: \underline{C} \times Ens \longrightarrow \underline{C}$ is the left adjoint of $\pitchfork: Ens^* \times \underline{C} \longrightarrow \underline{C}$ and this implies by Proposition 24, that $K_{\Gamma^{\#}}((\pitchfork \bullet (H^*_* \times I_{\underline{C}}))_{\underline{C} \times \underline{A}^*}) \dashv K_{\Gamma^{b}}((\sum \cdot (I_{\underline{C}} \times H))_{\underline{B}^* \times \underline{C}})$.

Université de Montréal

REFERENCES

1. M. André, "Categories of functors and adjoint functors", Am. J. of Math., vol. 88 (1966) 529-543.

2. P. Berthiaume, "On adjoints of functors between functor categories", Notices of the A.M.S., vol. 14 (1967), p. 708, note 67T-505.

3. ____________, "The functor evaluation", Notices of the A.M.S., vol. 15 (1968), p. 523, note 656-70.

4. J. R. Isbell, "Adequate subcategories", Illinois J. of Math., vol. 4 (1960), 541-552.

5. D. M. Kan, "Adjoint functors", T.A.M.S., vol. 87 (1958), 294-329.

6. J. Lambek, "Completions of categories", Springer Lecture Notes (1966).

7. F. W. Lawvere, "Functorial semantics of algebraic theories", thesis, Columbia (1963).

8. ____________, "The category of categories as a foundation of mathematics", Proceedings of the conference on categorical algebra, La Jolla 1965, Springer-Verlag. New York Inc.

9. F. E. J. Linton, "Autonomous categories and duality of functors", J. of Algebra, vol. 2 (1965), 315-349.

10. S. Mac Lane, "Categorical algebra", Bulletin of the A.M.S., vol. 71 (1965), 40-106.

11. B. Mitchell, "Theory of categories", Academic Press (1965).

12. F. Ulmer, "Properties of dense and relative adjoint functors", J. of Algebra, vol. 8 (1968), 77-95.

13. ________, "Representable functors with values in arbitrary categories", J. of Algebra, vol. 8 (1968), 96-129.

14. ________, "On Kan functor extensions", Mimeographed notes, ETH. Zurich.

AN ALTERNATIVE APPROACH TO UNIVERSAL ALGEBRA

by

R. F. C. Walters
Received March 6, 1969

The method of triples for defining ranked varieties of algebras over Sets (see [4] pp. 20-21) uses information about all free algebras. With the following related construction we need information about only two of the free algebras in defining the variety.

1. Definition of the construction.

Let $\underline{A}$ be a category. A device D over $\underline{A}$ consists of three things: X, η and E. X is a subclass of the objects of $\underline{A}$; η assigns to each $x \in X$ a morphism $\eta_x: x \longrightarrow Tx$ of $\underline{A}$ (and Tx will always denote the codomain of η_x); $E = \{E_{x,y};\ x, y \in X\}$ is a family of sets of morphisms where all the morphisms in $E_{x,y}$ have domain Tx and codomain Ty. We require the following additional properties:

(1) $\quad E_{y,z}E_{x,y} \subseteq E_{x,z} \qquad (x, y, z \in X)$,

(2) $\quad 1_{Tx} \in E_{x,x} \qquad (x \in X)$,

and (3) to each $\mu : x \longrightarrow Ty$ $(x, y \in X)$ there is a unique $\varepsilon \in E_{x,y}$ such that $\mu = \varepsilon\eta_x$. (For our theorem we shall be interested in devices over Sets with X of the form $\{x, \emptyset\}$.) We next define a category $\underline{A}^D$ of "D-objects and D-morphisms".

D-objects are pairs (a, F) where a is an object of $\underline{A}$ and $F = \{F_x;\ x \in X\}$ is a family of sets of morphisms of $\underline{A}$ where all morphisms in F_x have domain Tx and codomain a. We require further that

(1) $F_y E_{x,y} \subseteq F_x \quad (x, y \in X)$,

and (2) for each $x \in X$ and each $\mu : x \to a$ there is a unique $\varphi \in F_x$ such that $\mu = \varphi\eta_x$. A D-morphism from (a, F) to (b, G) is a morphism α: a $\to$ b such that $\alpha F_x \subseteq G_x$ $(x \in X)$. D-objects and D-morphisms make up the category $\underline{A}^D$ and there is an obvious forgetful functor U^D: $\underline{A}^D \to \underline{\text{Sets}}$ defined by:

U^D: (a, F) $\rightsquigarrow$ a

and U^D: (α:(a, F) $\to$ (b, G)) $\rightsquigarrow$ (α: a $\to$ b).

Every triple (T, η, μ) yields a device with X the objects of A, η_x: x $\to$ Tx the value of the natural transformation η at x, and $E_{x,y} = \{\mu(y)T(\alpha);\ \alpha: x \to Ty\}$. Further, all devices with X equal to the objects of $\underline{A}$ can be obtained from triples in this way and the construction of $\underline{A}^D$ is equivalent to the Eilenberg-Moore construction ([2]).

2. The classical definition of variety (for details see [1] or [5]).

To define algebras we need a set Ω of "formal operations" with a set $n(\omega)$ assigned to each $\omega \in \Omega$ called the "arity" of ω.

(In [1] $n(\omega)$ is always a finite ordinal and in [5] an ordinal.) Then an <u>Ω-algebra</u> $\underline{a}$ is a set a and to each $\omega \in \Omega$ an operation $\omega_{\underline{a}}: a^{n(\omega)} \longrightarrow a$. If $\alpha: n(\omega) \longrightarrow a$ we denote the image of α under $\omega_{\underline{a}}$ by $\omega_{\underline{a}}[\alpha]$. A <u>homomorphism</u> from $\underline{a}$ to $\underline{b}$ is a map $\lambda: a \longrightarrow b$ such that for all $\omega \in \Omega$ and all $\alpha: n(\omega) \longrightarrow a$ we have

$$\lambda\omega_{\underline{a}}[\alpha] = \omega_{\underline{b}}[\lambda\alpha].$$

All Ω-algebras and all homomorphisms constitute a category $\underline{\Omega\text{-Alg}}$. There is a clear forgetful functor $U_\Omega: \underline{\Omega\text{-Alg}} \longrightarrow \underline{\text{Sets}}$ which has a left adjoint W_Ω.

Now an Ω-law in variables x is a pair of elements of $U_\Omega W_\Omega x$. An Ω-algebra $\underline{a}$ satisfies the law (t_1, t_2) if $\alpha t_1 = \alpha t_2$ for every homomorphism $\alpha: W_\Omega x \longrightarrow \underline{a}$. Given L a set of Ω-laws in variables x, $\underline{V} = \underline{\text{Var (L)}}$ is the category of all Ω-algebras satisfying these laws (together with all homomorphisms). Again the natural forgetful functor $U_V: \underline{V} \longrightarrow \underline{\text{Sets}}$ has a left adjoint. (These categories of algebras <u>Var (L)</u> are called varieties.) Finally, a varietal functor U is a functor from some category $\underline{A}$ to <u>Sets</u> such that there exists a category $\underline{V} = \underline{\text{Var (L)}}$ and an isomorphism $K: \underline{V} \longrightarrow \underline{A}$ such that $UK = U_V$.

I am assuming that it is these varietal functors one studies in universal algebra rather than the particular way of constructing them. Certainly one can retrieve from a functor the "theory" of the corresponding variety (see [3]). However, we do not distinguish between two varieties with the same theory.

3. Theorem.

If $D = (X, \eta, E)$ is a device over Sets with $X = \{x, \emptyset\}$, then U^D: $\underline{\text{Sets}}^D \longrightarrow \underline{\text{Sets}}$ is a varietal functor. Further, to each varietal functor $U: \underline{A} \longrightarrow \underline{\text{Sets}}$ there is a device D of this sort and an isomorphism $K: \underline{A} \longrightarrow \underline{\text{Sets}}^D$ such that $U^D K = U$.

Proof.

(i) Given $D = (X, \eta, E)$ with $X = \{x, \emptyset\}$;

to construct the required variety we need to select an operation set Ω, an arity function n, a morphism $\xi: \Omega \longrightarrow Tx$ and to each $\omega \in \Omega$ an injection $\iota_\omega : n(\omega) \longrightarrow x$. The following is a suitable selection: $\Omega = Tx$; $n(\omega) = \emptyset$ if $\omega \in$ image $(\varepsilon_{\emptyset,x})$ ($\varepsilon_{\emptyset,x}$ is the single morphism in $E_{\emptyset,x}$) and $n(\omega) = x$ for all other $\omega \in \Omega$; $\xi = 1_{Tx}$; $\iota_\omega = \emptyset \longrightarrow x$ if $n(\omega) = \emptyset$ and $\iota_\omega = 1_x$ if $n(\omega) = x$. More generally, any choice which satisfies the following properties will suffice:

(1) $n(\omega) = \emptyset$ only if $\xi\omega \in$ image $(\varepsilon_{\emptyset,x})$,

(2) $n(\omega) = y \neq \emptyset$ implies that for any $\alpha_1, \alpha_2: x \longrightarrow Tx$

$$\alpha_1 \iota_\omega = \alpha_2 \iota_\omega \Rightarrow \varepsilon_1 \xi\omega = \varepsilon_2 \xi\omega$$

where $\alpha_i = \varepsilon_i \eta_x \quad (\varepsilon_i \in E_{x,x};\ i = 1, 2)$,

and (3) if $\underline{Tx} = \text{alg}(Tx, \{E_{x,x}, E_{\emptyset,x}\})$ and

$\underline{T\emptyset} = \text{alg}(T\emptyset, \{E_{x,\emptyset}, E_{\emptyset,\emptyset}\})$ are defined as in the next paragraph, (the images of) η_x: x $\to$ Tx and $\eta_\emptyset$: $\emptyset \to$ T$\emptyset$ generate (in the algebraic sense) <u>Tx</u> and <u>T$\emptyset$</u> respectively.

Now given a selection of Ω and n and associated morphisms ξ and ι_ω ($\omega \in \Omega$) satisfying (1) and (2) above, we define to each (a, F) $\in$ <u>Sets</u>^{D} an Ω-algebra alg(a, F) as follows. alg(a, F) has underlying set a and if n(ω) = y and α: y $\to$ a then $\alpha = \beta\iota_\omega$ for some β and $\beta = \varphi\eta_x(\varphi \in F_x)$. (A suitable morphism β may not exist when n(ω) = $\emptyset$ and a = $\emptyset$ but this situation cannot occur since the existence of a nullary operation ω implies that T$\emptyset \neq \emptyset$ and hence, since $F_\emptyset$ is non-empty, that a $\neq \emptyset$.) We put $\omega_{\underline{a}}[\alpha] = \varphi\xi\omega$. We have to check that $\omega_{\underline{a}}[\alpha]$ does not depend on the particular β chosen; that is, that $\beta_1\iota_\omega = \beta_2\iota_\omega$ implies that $\varphi_1\xi\omega = \varphi_2\xi\omega$ where $\varphi_i\eta_x = \beta_i$ (i = 1, 2) ($\varphi_i \in F_x$). When n(ω) = $\emptyset$ we use the fact that $\varphi_i \cdot \varepsilon_{\emptyset,x} = \varphi_\emptyset$ ($\varepsilon_{\emptyset,x} \in E_{\emptyset,x}$; $\varphi_\emptyset \in F_\emptyset$). When n($\omega$) = y $\neq \emptyset$ there certainly exists a map β : x $\to$ a and a map γ: x $\to$ x such that $\beta = \beta_i\gamma$ (i = 1, 2) and $\gamma\iota_\omega = \iota_\omega$. Let $\eta_x\gamma = \varepsilon\eta_x$ ($\varepsilon \in E_{x,x}$) and $\beta = \varphi\eta_x$ ($\varphi \in F_x$). Then $(\varphi_i\varepsilon)\eta_x = \varphi_i\eta_x\gamma = \beta$ and hence $\varphi_i\varepsilon = \varphi$. Now since $\eta_x\gamma\iota_\omega = \eta_x\iota_\omega$ we have $\varepsilon\xi\omega = \xi\omega$. (This is restriction (2) on the selection of Ω and n.) Hence $\varphi_1\xi\omega = \varphi_1\varepsilon\xi\omega = \varphi\xi\omega = \varphi_2\varepsilon\xi\omega = \varphi_2\xi\omega$, which is what we wished to prove. Thus alg(a, F) is a well defined Ω-algebra.

(ii) Next we wish to show that λ: a $\to$ b is a homomorphism from $\underline{a}$ = alg (a, F) to $\underline{b}$ = alg (b, G) if and only if it is a D-morphism from (a, F) to (b, G).

Suppose λ is a D-morphism, then $\lambda F_x \subseteq G_x$. For operations ω of arity y consider $\alpha : y \longrightarrow a$. We need that

$$\lambda \omega_{\underline{a}}[\alpha] = \omega_{\underline{b}}[\lambda\alpha].$$

Let $\alpha = \varphi\eta_x \iota_\omega (\varphi \in F_x)$. Then the left hand side is equal to $\lambda\varphi\xi\omega$. Further $\lambda\alpha = \lambda\varphi\eta_x\iota_\omega$ and since $\lambda\varphi \in G_x$ the right hand side is also $\lambda\varphi\xi\omega$.

Conversely let λ be a homomorphism from $\underline{a}$ to $\underline{b}$. Consider $\lambda\varphi$ where $\varphi \in F_x$. There exists a $\gamma \in G_x$ such that $\gamma\eta_x = \lambda\varphi\eta_x$. Now $\lambda\varphi$ and γ are both homomorphisms from $\underline{Tx}$ to $\underline{b}$ and they agree on the generators so they are equal. That is, $\lambda F_x \subseteq G_x$. It is clear that $\lambda F_\emptyset = G_\emptyset$ since for any $\varphi \in F_x$, $\varphi\varepsilon_{\emptyset,x}$ is the only map in $F_\emptyset$ and $\lambda\varphi\varepsilon_{\emptyset,x}$ is the only map in $G_\emptyset$.

(iii) $\mathrm{alg}(a, F) = \mathrm{alg}(b, G) \Rightarrow (a, F) = (b, G)$.

Clearly the left hand side implies that a = b. Suppose that the right hand side is nevertheless false. Then either $F_\emptyset \neq G_\emptyset$ or $F_x \neq G_x$. In the first case it follows that $\varphi\varepsilon_{\emptyset,x} \neq \gamma\varepsilon_{\emptyset,x}$ and hence that $\varphi \neq \gamma$ for any $\varphi \in F_x$, $\gamma \in G_x$. Hence we may assume that $F_x \neq G_x$. Then there exists $\varphi \in F_x$ and $\gamma \in G_x$ such that $\varphi\eta_x = \gamma\eta_x$ but $\varphi \neq \gamma$. This however cannot occur since φ and γ are both homomorphisms from $\underline{Tx}$ to $\underline{a}$ (where $\underline{a} = \mathrm{alg}(a, F) = \mathrm{alg}(a, G)$) and they agree on the generators so they are equal.

(iv) We next wish to identify the algebras alg(a, F) as belonging to a certain variety. Let $\underline{Wx}$ be the free Ω-algebra with underlying set Wx, freely generated by $x \xrightarrow{\tau} Wx$. Then there is a unique homomorphism ν from Wx to Tx such that $\nu\tau = \eta_x$. We define a set of laws L as follows:

$$L = \{(t_1, t_2);\ t_1, t_2 \in Wx \text{ and } \nu t_1 = \nu t_2\}.$$

All algebras a = alg(a, F) satisfy these laws. Clearly this would follow if $F_x\nu$ were all homomorphisms from Wx to a. But to each $\mu : x \to a$ there is a homomorphism $\varphi\nu$ from Wx to a belonging to $F_x\nu$ such that $\varphi\nu\tau = \mu$. Any homomorphism from Wx to a must agree with one of these on the generators of Wx and hence must be one of them.

(v) Any algebra a of Var(L) is of the form alg(a, F) for some (a, F) $\in$ Sets^D. Take F_x to be all homomorphisms from Tx to a and $F_\emptyset$ to be all homomorphisms from T$\emptyset$ to a. Property (1) for D-objects is then clearly true for (a, F). To check (2) consider any $\mu : x \to a$. There exists a homomorphism $\lambda : Wx \to a$ such that $\lambda\tau = \mu$. Now ν is an epimorphism since $\eta_x : x \to Tx$ generates Tx. Further whenever $\nu t_1 = \nu t_2$ ($t_1, t_2 \in Wx$) then $(t_1, t_2) \in L$ so that $\lambda t_1 = \lambda t_2$. Under these conditions there exists a homomorphism $\kappa : Tx \to a$ such that $\kappa\nu = \lambda$ and hence $\kappa\eta_x = \kappa\nu\tau = \lambda\tau = \mu$. Since η_x generates Tx there is at most one such homomorphism. To check (2) we also have to show that any $\emptyset \to a$ is of the form $\varphi_\emptyset\eta_\emptyset$ for a unique $\varphi_\emptyset \in F_\emptyset$. This amounts to showing that $F_\emptyset$ contains precisely one element. It contains at most one since $\emptyset \to T\emptyset$ generates T$\emptyset$. If $T\emptyset = \emptyset$

there is the empty homomorphism from $\underline{T\emptyset}$ to $\underline{a}$. If $T\emptyset \neq \emptyset$ then there are nullary operations so that $\underline{a}$ cannot be the empty algebra and hence F_x is non-empty. Then $\varphi \varepsilon_{\emptyset,x} \in F_\emptyset$ for any $\varphi \in F_x$.

It remains to be shown that $\underline{\underline{a}} = \underline{a}$ where $\underline{\underline{a}} = \mathrm{alg}(a, F)$. Consider $\omega \in \Omega$ and $\alpha: n(\omega) \to a$. Let $\alpha = \varphi \eta_x \iota_\omega$ ($\varphi \in F_x$). Then $\omega_{\underline{\underline{a}}}[\alpha] = \varphi \xi \omega = \varphi \omega_{\underline{Tx}}[\eta_x \iota_\omega] = \omega_{\underline{a}}[\varphi \eta_x \iota_\omega] = \omega_{\underline{a}}[\alpha]$.

Thus we have shown that alg: $\underline{\mathrm{Sets}}^D \to \underline{\mathrm{Var}(L)}$ defined by:

$$\mathrm{alg}: (a, F) \rightsquigarrow \mathrm{alg}(a, F)$$

and
$$\mathrm{alg}: (\alpha : (a, F) \to (b, G)) \rightsquigarrow (\alpha : \mathrm{alg}(a, F) \to \mathrm{alg}(b, G))$$

is an isomorphism and it is clear that $U_V \, \mathrm{alg} = U^D$. Hence U^D is a varietal functor.

(vi) We shall now discuss the second part of the theorem. Let $\underline{V}$ be any variety with operation set Ω and arity function n. Let $\underline{Wx}$ be the free Ω-algebra freely generated by $\uparrow : x \to Wx$ and let $\underline{Tu}$ be the $\underline{V}$-free algebra freely generated by $\eta_u: u \to Tu$. Let ν be the homomorphism from $\underline{Wx}$ to $\underline{Tx}$ such that $\nu\uparrow = \eta_x$. Then it is a fact of universal algebra that for all sufficiently large x, $\underline{V} = \underline{\mathrm{Var}(L)}$ where

$$L = \{(t_1, t_2);\ t_1, t_2 \in Wx \text{ and } \nu t_1 = \nu t_2\}.$$

Take such an x with $|x| > |n(\omega)|$ for all $\omega \in \Omega$. Then consider the device with $X = \{x, \emptyset\}$, η_x and $\eta_\emptyset$ as above, and $E_{u,v}$ all homomorphisms from $\underline{Tu}$ to $\underline{Tv}$ ($u, v \in X$). Certainly $D = (X, \eta, E)$

is a device. We wish to consider the variety $\underline{V}'$ obtained from this device by the method given in the earlier parts of this theorem. Now Ω and n form a suitable operation set and arity function for $\underline{V}'$, if ι_ω is taken to be any injection from $n(\omega)$ to x and ξ is defined by:

$$\xi : \omega \rightsquigarrow \omega_{\underline{\underline{Tx}}}[\eta_x \iota_\omega].$$

Now for any $\alpha : n(\omega) \longrightarrow Tx$ let $\alpha = \varepsilon \eta_x \iota_\omega$ $(\varepsilon \in E_{x,x})$. Then if $\underline{\underline{Tx}} = \text{alg}(Tx, \{E_{x,x}, E_{\emptyset,x}\})$ we see that

$$\omega_{\underline{\underline{Tx}}}[\alpha] = \varepsilon \xi \omega = \varepsilon \omega_{\underline{Tx}}[\eta_x \iota_\omega] = \omega_{\underline{Tx}}[\alpha].$$

This means that $\underline{\underline{Tx}} = \underline{Tx}$. Now $\underline{V}'$ and $\underline{V}$ have the same operation set and arity function. Further the laws defining $\underline{V}'$ are obtained from η_x and $\underline{\underline{Tx}}$ in precisely the same way that the laws L of $\underline{V}$ are obtained from η_x and $\underline{Tx}$. Hence $\underline{V}' = \underline{V}$ and $U_{V'} = U_V$. So there exists an isomorphism $K: \underline{\text{Sets}}^D \longrightarrow \underline{V}$ such that $U_V K = U^D$, and this is what we were required to prove.

REFERENCES

[1] P. M. Cohn: Universal Algebra, Harper and Row, New York (1965).

[2] S. Eilenberg and J. C. Moore: Adjoint functors and triples; Ill. J. Math. 9, 381-398 (1965).

[3] F. E. J. Linton: Some aspects of equational categories; Proceedings of the conference on categorical algebra (La Jolla, 1965), Springer, Berlin (1966).

[4] E. G. Manes: A triple miscellany, Dissertation, Wesleyan University, Middletown, Conn. (1967).

[5] J. Słomiński: The theory of abstract algebras with infinitary operations, Rozprawy Mat. vol. 18, Warsaw (1959).

Australian National University,

Canberra.

VARIATIONS ON BECK'S TRIPLEABILITY CRITERION

by

J. DUSKIN*

Received November 3, 1968

(0.1) Introduction. Beck's criterion is a convenient condition on a functor U: $\mathcal{A} \to \mathcal{B}$ whose verification (in the presence of a left-adjoint) is necessary and sufficient for the category $\mathcal{A}$ to be equivalent to the category of Eilenberg-Moore algebras defined by the ordered pair of adjoint functors (U,F).

It will be the point of this article to show that under very weak additional hypotheses, Beck's criterion (the condition (B) of (3.0)) may be replaced with an entirely analogous condition ((B*) of (3.2)) which involves only the U-contractile equivalence pairs of $\mathcal{A}$ rather than arbitrary U-contractile pairs. The advantage of this latter is, of course, that one normally has more information about "passage to the quotient" by equivalence relations than by ordinary double arrows in most concrete situations. The proof of this result is based on certain properties of contractile pairs which are resumed in section 2 and although somewhat complicated technically, is simple in outline.

---

* The research for this article was done while the author was a National Research Council of Canada Overseas Post-doctoral Fellow at the Universities of Strasbourg and Paris during 1966-68. The results of section 5 were presented in 1967 before a seminar on the theory of categories at the University of Paris directed by J. Bénabou as well as before the Oberwolfach Conference on Categorical and Universal Algebra in July of 1968.

Having established this result, we apply it to several situations: In section 4 to give variations of Linton's theorem characterizing categories which are tripleable over certain "set like" categories: (4.2) and (4.4) and in section 5 to give various necessary and sufficient conditions for a category to be tripleable over a functor category (5.11). In particular, this latter allows us to characterize internally those categories which are tripleable over (ENS) by means of a particular projective generator (5.13) as well as those whose opposite category are tripleable over (ENS) (5.15) in a manner entirely analogous to Lawvere (1963). As a number of classical theorems may be viewed as key portions of the verification of these conditions, a full discussion of this and its relation with Gabriel retracts will be given in a separate article, as well as a full discussion of Giraud's characterization theorem for categories of sheaves of sets ($\mathcal{U}$-topos) which is mentioned in (5.17). Equivalence with functor categories is discussed in section 6 as a special case of this latter theorem. We conclude with a general discussion of the existence of co-limits and certain left-adjoints in section 7. The reader is cautioned that the usage of the terms "exact" and "co-exact" is at variance with that of some authors. The notions of equivalence pair and effectivity as used here are due to Grothendieck (1961) and differ slightly from Lawvere's (1963) classification where an equivalence pair is called a pre-congruence and an effective equivalence pair is called a congruence.

Their introduction into contexts such as ours is of course due to Lawvere.

(0.2) <u>Notations and conventions</u>. In this article all categories are supposed $\mathcal{U}$-categories for $\mathcal{U}$ some universe which remains fixed during the discussion. If $\mathcal{A}$ is a category, we will denote the set $\mathrm{Hom}_{\mathcal{A}}(T,X)$ of all arrows in $\mathcal{A}$ with domain T and co-domain X by $\mathcal{A}(T,X)$. If $f\colon X \to Y$ is an arrow in $\mathcal{A}$, the canonical mapping of $\mathcal{A}(T,X)$ into $\mathcal{A}(T,Y)$ will be denoted by $\mathcal{A}(T,f)$ while that of $\mathcal{A}(Y,T)$ into $\mathcal{A}(X,T)$ by $\mathcal{A}(f,T)$. If $f\colon X \to Y$ and $g\colon X \to Z$ are mappings, we will write $\langle f,g\rangle : X \to Y \times Z$ for the mapping defined by $\langle\langle x \mapsto (f(x),g(x))\rangle\rangle$ and will follow this same notation for those mappings having the same graph as $\langle f,g\rangle$. A diagram of sets of the form $\langle\langle X \xrightarrow{f} Y \underset{h}{\overset{g}{\rightrightarrows}} Z\rangle\rangle$ will be said to be a <u>complex</u> provided gf = hf, and called <u>exact</u> if gh = hf, and in addition the mapping f defines a bijection of X onto the subset Ker(g,h) of Y whose elements are all those y such that g(y) = h(y). A diagram of the same type in $\mathcal{A}$ will be called <u>exact</u> if, for any $T \in \mathcal{O}b(\mathcal{A})$, the diagram

$$\langle\langle \mathcal{A}(T,X) \xrightarrow{\mathcal{A}(T,f)} \mathcal{A}(T,Y) \underset{\mathcal{A}(T,h)}{\overset{\mathcal{A}(T,g)}{\rightrightarrows}} \mathcal{A}(T,Z)\rangle\rangle$$

of sets and mappings is exact. A diagram of the form $X \underset{h}{\overset{g}{\rightrightarrows}} Y \xrightarrow{f} Z$ in any category will be called <u>exact</u> provided for all $T \in \mathcal{O}b(\mathcal{A})$ the diagram of sets

$$\langle\langle \mathcal{A}(T,X) \xrightarrow{\langle \mathcal{A}(T,g),\,\mathcal{A}(T,h)\rangle} \mathcal{A}(T,Y) \times \mathcal{A}(T,Y) \underset{\mathcal{A}(T,f)pr_2}{\overset{\mathcal{A}(T,f)pr_1}{\rightrightarrows}} \mathcal{A}(T,Z)\rangle\rangle$$

is exact, and <u>co-exact</u> provided for all $T \in \mathcal{O}b(\mathcal{A})$,

$$\langle\langle \mathcal{A}(Z,T) \xrightarrow{\mathcal{A}(f,T)} \mathcal{A}(Y,T) \underset{\mathcal{A}(h,T)}{\overset{\mathcal{A}(g,T)}{\rightrightarrows}} \mathcal{A}(X,T)\rangle\rangle$$

is exact, i.e.,

$X \rightrightarrows Y \longrightarrow Z$ is exact in $\mathcal{A}^{op}$. If it is both exact and co-exact it will be called bi-exact. A square

$$\begin{array}{ccc} A & \xrightarrow{f} & C \\ {\scriptstyle a}\downarrow & & \downarrow{\scriptstyle b} \\ B & \xrightarrow[g]{} & D \end{array}$$

is called cartesian (or a pull-back diagram) if for all $T \in \mathcal{O}b(\mathcal{A})$, the diagram

$$\mathcal{A}(T,A) \xrightarrow{\langle \mathcal{A}(T,a),\, \mathcal{A}(T,f)\rangle} \mathcal{A}(T,B) \times \mathcal{A}(T,C) \overset{\mathcal{A}(T,g)pr_1}{\underset{\mathcal{A}(T,b)pr_2}{\rightrightarrows}} \mathcal{A}(T,D)$$

is exact.

We now use these definitions to state two elementary lemmas involving exact diagrams in the category of sets. The proofs are trivial and are left to the reader, as is also the multitude of corollaries which may be drawn from them in arbitrary categories using the above definitions.

(0.3) Lemma. Let

$$\begin{array}{ccccc} \mathfrak{X}: & X_1 & \xrightarrow{x_1} & X_2 & \overset{x_2}{\underset{x_3}{\rightrightarrows}} & X_3 \\ & \xi_1 \downarrow & (D) & \downarrow \xi_2 & & \downarrow \xi_3 \\ \mathcal{Y}: & Y_1 & \xrightarrow{y_1} & Y_2 & \overset{y_2}{\underset{y_3}{\rightrightarrows}} & Y_3 \end{array}$$

be a diagram of sets and applications such that $\mathfrak{X}$ and $\mathcal{Y}$ are complexes (i.e. $x_2x_1 = x_3x_1$ and $y_2y_1 = y_3y_1$) and whose component squares are sequentially commutative (i.e. $y_1\xi_1 = \xi_2x_1$, $y_2\xi_2 = \xi_3x_2$, $y_3\xi_2 = \xi_3x_3$). Then

(a) if the square (D) is cartesian, $\mathcal{Y}$ exact $\Rightarrow \mathfrak{X}$ exact, and

(b) $\mathfrak{X}$ exact, $\mathcal{Y}$ exact and ξ_3 an injection $\Rightarrow$ (D) is cartesian.

(0.4) Corollary. Let

$$\begin{array}{lccccc}
\mathfrak{X}: & X_1 & \xrightarrow{x_1} & X_2 & \overset{x_2}{\underset{x_3}{\rightrightarrows}} & X_3 \\
 & \xi_1 \downarrow \ (D) & & \xi_2 \downarrow & & \xi_3 \downarrow \\
\mathcal{Y}: & Y_1 & \xrightarrow{y_1} & Y_2 & \overset{y_2}{\underset{y_3}{\rightrightarrows}} & Y_3 \\
 & \rho_1^1 \downdownarrows \rho_2^1 & & \rho_1^2 \downdownarrows \rho_2^2 & & \rho_1^3 \downdownarrows \rho_2^3 \\
\mathfrak{Z}: & Z_1 & \xrightarrow{z_1} & Z_2 & \overset{z_2}{\underset{z_3}{\rightrightarrows}} & Z_3 \\
 & A & & B & & C
\end{array}$$

be a diagram of sets and mappings of "bi-simplicial type" (i.e. here it will suffice that the two component diagrams involving the square D and the respective arrows ξ_3 and z_1 satisfy the hypothesis of (0.3)) in which the columns A, B, C are exact. Then

$$\mathcal{Y} \text{ exact and } \mathfrak{Z} \text{ exact} \Rightarrow \mathfrak{X} \text{ exact.}$$

In effect $\mathfrak{Z}$ exact $\Rightarrow z_1$ is an injection and then the exactness of A and B implies that D is cartesian; but then $\mathcal{Y}$ exact $\Rightarrow \mathfrak{X}$ exact. Q.E.D.

In what follows, the set of arrows of a category $\mathcal{C}$ will be abbreviated as $\mathcal{A}r(\mathcal{C})$, and the set of objects of $\mathcal{C}$ by $\mathcal{O}b(\mathcal{C})$. A category is called ($\mathcal{U}$-) small provided both the sets $\mathcal{A}r(\mathcal{C})$ and $\mathcal{O}b(\mathcal{C})$ are equipotent with sets which are elements of the universe $\mathcal{U}$.

1. Equivalence pairs in categories

(1.0) Definition. Let $X_1 \overset{x_1}{\underset{x_2}{\rightrightarrows}} X_2$ be a double arrow in a category $\mathcal{A}$.

We will say that (x_1,x_2) is an <u>equivalence pair</u> if, for all $T \in \mathcal{O}b(\mathcal{A})$, the lifted set mapping

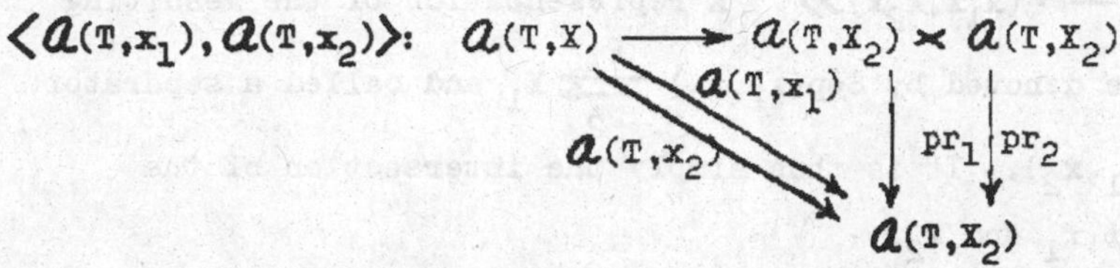

defined by $\langle\langle x \longmapsto (x_1x, x_2x) \rangle\rangle$ is an injection which has as its image the graph of an equivalence relation on the set $\mathcal{A}(T,X_2)$.

(1.1) For example, if $f: X \longrightarrow Y$ is an arrow in $\mathcal{A}$, then for each $T \in \mathcal{O}b(\mathcal{A})$ the fiber product $\mathcal{A}(T,X) \underset{\mathcal{A}(T,f),\mathcal{A}(T,f)}{\times} \mathcal{A}(T,X)$ is the graph of the equivalence relation $\langle\langle \mathcal{A}(T,f)(x) = \mathcal{A}(T,f)(y) \rangle\rangle$ associated with the mapping $\mathcal{A}(T,f)$. If the functor defined by $\langle\langle T \longmapsto \mathcal{A}(T,X) \underset{\mathcal{A}(T,f),\mathcal{A}(T,f)}{\times} \mathcal{A}(T,X) \rangle\rangle$ is representable, we will say that an <u>equivalence pair</u> (or <u>kernel pair</u>) <u>associated with</u> f exists and will use the notation $\mathcal{R}_{eq}(f) \overset{p_1}{\underset{p_2}{\rightrightarrows}} X$ (or $\mathcal{R}(f) \overset{p_1}{\underset{p_2}{\rightrightarrows}} X$) to indicate the resulting fiber product $X \underset{f,f}{\times} X \overset{p_1}{\underset{p_2}{\rightrightarrows}} X$. If for any $f \in \mathcal{A}r(\mathcal{A})$, an equivalence pair associated with f exists we will say that the category $\mathcal{A}$ has <u>square fiber products</u>.

(1.2) In a similar fashion, we say that a category $\mathcal{A}$ admits <u>separators of pairs of the</u> form (x_1,x_2): $X_1 \rightrightarrows X_2$ if, for each such double arrow, the functor defined by $\langle\langle T \longmapsto \mathcal{R}(\langle \mathcal{A}(T,x_1), \mathcal{A}(T,x_2) \rangle) \rangle\rangle$ is representable. Here $\mathcal{R}(\langle \mathcal{A}(T,x_1), \mathcal{A}(T,x_2) \rangle)$

designates the graph of the equivalence relation associated with the mapping $\langle \mathcal{A}(T,x_1), \mathcal{A}(T,x_2)\rangle : \mathcal{A}(T,X_1) \longrightarrow \mathcal{A}(T,X_2) \times \mathcal{A}(T,X_2)$ given by $\langle\langle x \longmapsto (x_1 x, x_2 x)\rangle\rangle$. A representation of the resulting functor will be denoted by $\mathrm{Sep}(x_1,x_2) \underset{p_2}{\overset{p_1}{\rightrightarrows}} X_1$ and called a separator of the pair (x_1,x_2). It is thus simply the intersection of the kernel pairs of x_1 and x_2.

(1.3) It is evident that if $\mathcal{A}$ admits square fiber products and if the product of an object with itself always exists (or if $\mathcal{A}$ has a final object), then $\mathcal{A}$ has such separators.

$$\begin{array}{ccccc} & & & & X_2 \times X_2 \\ & & & \nearrow^{\langle x_1,x_2\rangle} & \downarrow\downarrow \mathrm{pr}_1\ \mathrm{pr}_2 \\ \mathrm{Sep}(x_1,x_2) = \mathcal{K}(\langle x_1,x_2\rangle) & \rightrightarrows & X_1 & \underset{x_2}{\overset{x_1}{\rightrightarrows}} & X_2 \end{array}$$

This is also equally the case if $\mathcal{A}$ admits fiber products and kernels of pairs of arrows.

Note that if $\mathcal{A}$ has square fiber products and if the pair in question $X_1 \underset{x_2}{\overset{x_1}{\rightrightarrows}} X_2$ has an arrow $X_2 \xrightarrow{x_3} X_3$ such that $x_3 x_1 = x_3 x_2$ (for example, if a co-kernel of the pair (x_1,x_2) exists) then a separator of (x_1,x_2) exists, it being then given by $\mathcal{K}(\langle x_1,x_2\rangle)$ where $\langle x_1,x_2\rangle : X_1 \longrightarrow \mathcal{K}(x_3)$ is that unique arrow such that $p_1(\langle x_1,x_2\rangle) = x_1$ and $p_2(\langle x_1,x_2\rangle) = x_2$ which exists since $x_3 x_1 = x_3 x_2$.

It is clear that if a category admits separators, then it admits square fiber products since one will have a canonical isomorphism of $\mathcal{R}(f)$ with Sep(f,f).

(1.4) <u>Definition</u>. An <u>equivalence pair</u> $X_1 \overset{x_1}{\underset{x_2}{\rightrightarrows}} X_2$ <u>is effective</u> if a co-kernel $v\colon X_2 \to \mathrm{Coker}(x_1,x_2)$ exists and the resulting square

$$\begin{array}{ccc} X_1 & \xrightarrow{x_1} & X_2 \\ {\scriptstyle x_2}\downarrow & & \downarrow{\scriptstyle v} \\ X_2 & \xrightarrow{v} & \mathrm{Coker}(x_1,x_2) \end{array}$$

is cartesian (i.e. $\mathcal{R}(v) \rightrightarrows X_2$ exists and $\langle x_1,x_2\rangle : X_1 \to \mathcal{R}(v)$ is an isomorphism). The square in question is then bi-cartesian (at once cartesian and co-cartesian).

It is trivial to prove that in order that an equivalence pair associated with some arrow (i.e. a kernel pair) be effective it is necessary and sufficient that a co-kernel of the couple exist. (In general we will call any double arrow $R \rightrightarrows S$ <u>semi-effective</u> if it has a co-kernel). The same is true of those equivalence pairs which arise as separators and, more generally, with any member of a class of equivalence pairs which we will not define until later (7.5) the class of <u>strict</u> equivalence pairs.

The «galois dual» of the preceding definition is the following: An <u>arrow</u> $f\colon X_1 \to X_2$ is said to be an <u>effective</u> epimorphism if the fiber product $X_1 \underset{f,f}{\times} X_1 \overset{d_0}{\underset{d_1}{\rightrightarrows}} X_1$ exists and f is a co-kernel of (d_0,d_1).

In a functor category $\langle\langle R \rightrightarrows S$ is an equivalence pair$\rangle\rangle$ $\Leftrightarrow$ $\langle\langle$for all $T, R(T) \rightrightarrows S(T)$ is an equivalence pair$\rangle\rangle$.

2. Contractile pairs

We first make some remarks about contractile pairs in the category of sets (ENS) which may be immediately extended to any category having square fiber products by the usual methods.

(2.0) In what follows, let $(x_1,x_2,m) : X_1 \rightleftarrows X_2$ be a contractile pair (i.e. an ordered pair of mappings (x_1,x_2): $X_1 \rightrightarrows X_2$ supplied with an additional mapping m: $X_2 \longrightarrow X_1$, called a contraction, which has the property that (a) $x_1m = id(X_2)$ and (b) $x_2mx_1 = x_2mx_2$.

(2.1) If $\mathcal{R}(xm)$ is the set of pairs $(z,v) \in X_2 \times X_2$ such that $x_2m(z) = x_2m(w)$, then the equivalence pair (pr_1,pr_2): $\mathcal{R}(x_2m) \rightrightarrows X_2$ is a contractile pair when it is supplied with the contraction m^*: $X_2 \longrightarrow \mathcal{R}(x_2m)$ given by the assignment $\langle\langle z \longmapsto (x_1m(z),x_2m(z)) = (z,x_2m(z))\rangle\rangle$.

This mapping is defined since, by definition, one has $x_2m(z) = x_2mx_2(m)(z)$ for any $z \in X_2$. Moreover $pr_1m^* = id$ and $x_2mpr_1 = x_2mpr_2$, and since $pr_2m^* = x_2m$, one thus has $(pr_2m^*)pr_1 = (pr_2m^*)pr_2$ as desired.

(2.2) Let $(x,y) \in X_1 \times X_1$ be an ordered pair such that $x_2(x) = x_2(y)$; by definition one then has $x_2mx_1(x) = x_2mx_2(x) = x_2mx_2(y) = x_2mx_1(y)$. Thus there exists a mapping $x_1 \hat{\times} x_1$: $\mathcal{R}(x_2) \longrightarrow \mathcal{R}(x_2m)$ defined by $\langle\langle(x,y) \longmapsto (x_1(x),x_1(y))\rangle\rangle$. This mapping $x_1 \hat{\times} x_1$ admits a section, defined by $\langle\langle (z,w) \mapsto (m(z),m(w)) \rangle\rangle$, since the relation $\langle\langle (z,w) \in \mathcal{R}(x_2 m) \rangle\rangle$ is equivalent to $\langle\langle x_2 m(z) = x_2 m(w) \rangle\rangle$, that is to say to $\langle\langle (m(z),m(w)) \in \mathcal{R}(x_2) \rangle\rangle$, and finally $(x_1(m(z)),x_1(m(w))) = (z,w)$.

Now if $f\colon X \to Y$ is a mapping which admits a section $s\colon Y \to X$, then $\mathcal{R}(f) \overset{pr_1}{\underset{pr_2}{\rightrightarrows}} X$ is a contractile equivalence pair when it is supplied with the contraction $\mathfrak{T}\colon X \to \mathcal{R}(f)$ given by $\langle\langle x \mapsto (x, sf(x)) \rangle\rangle$ since $(sf)(sf) = sf$. (Conversely, if $X_1 \overset{x_1}{\underset{x_2}{\rightrightarrows}} X_2$ is a contractile equivalence pair, its quotient $p\colon X_2 \to Q$ admits a section $s\colon Q \to X_2$ defined by that unique mapping s such that $sp = x_2 m$.)

As a result we have that $\mathcal{R}(x_1 \hat{\times} x_1) \overset{pr_1}{\underset{pr_2}{\rightrightarrows}} \mathcal{R}(x_2)$ is <u>a contractile, effective, equivalence pair with co-kernel</u>, $x_1 \hat{\times} x_1\colon \mathcal{R}(x_2) \to \mathcal{R}(x_2 m)$.

(2.3) If $q_1, q_2\colon \mathcal{R}(x_2) \rightrightarrows X_1$ are the canonical projections, the equivalence relation associated with the mapping $x_1 \hat{\times} x_1$ is equivalent to that of the mapping $\langle x_1 q_1, x_1 q_2 \rangle\colon \mathcal{R}(x_2) \to X_2 \times X_2$ defined by $\langle\langle (x,y) \mapsto (x_1(x), x_1(y)) \rangle\rangle$. <u>The separator of the pair</u> $(x_1 q_1, x_1 q_2)\colon \mathcal{R}(x_2) \rightrightarrows X_2$ <u>is thus the same as the equivalence pair associated</u> with the mapping $x_1 \hat{\times} x_1$, i.e. $\mathrm{Sep}(x_1 q_1, x_1 q_2) = \mathcal{R}(x_1 \hat{\times} x_1)$ and $\mathcal{R}(x_2 m)$ is bijectively equivalent to the image of $\mathcal{R}(x_2)$ in $X_2 \times X_2$ under the mapping $x_1 \hat{\times} x_1$.

The preceding remarks give rise to the following proposition:

(2.4) Proposition. Let $(x_1, x_2, m): X_1 \rightleftarrows X_2$ be a contractile pair in a category $\mathcal{A}$ in which the equivalence pairs $\mathcal{R}(x_2 m) \overset{p_1}{\underset{p_2}{\rightrightarrows}} X_2$ and $\mathcal{R}(x_2) \overset{q_1}{\underset{q_2}{\rightrightarrows}} X_1$ exist. Then

1° the pair (p_1, p_2) is a contractile equivalence pair with contraction defined by $\langle id(X_2), x_2 m \rangle$;

2° the arrow $x_1 \hat{\asymp} x_1 : \mathcal{R}(x_2) \rightarrow \mathcal{R}(x_2 m)$ defined by $x_1 \hat{\asymp} x_1 = \langle x_1 q_1, x_1 q_2 \rangle$ is a retraction with section s defined by $s = \langle m p_1, m p_2 \rangle$;

3° An equivalence pair associated with the arrow $x_1 \hat{\asymp} x_1$ exists if, and only if, a separator of the pair $(x_1 q_1, x_1 q_2)$ exists, in which case they are equivalence pairs contractile, effective, and canonically isomorphic.

Thus in such a category a contractile pair gives rise to a diagram

$$\begin{array}{ccccccc}
\mathcal{R}(x_1 \hat{\asymp} x_1) & \xrightarrow{\sim} & Sep(x_1 q_1, x_1 q_2) & \overset{r_1}{\underset{r_2}{\rightleftarrows}} & \mathcal{R}(x_2) & \overset{x_1 \hat{\asymp} x_1}{\rightleftarrows} & \mathcal{R}(x_2 m) \\
 & & & & q_1 \downarrow\downarrow q_2 & \nearrow \langle x_1, x_2 \rangle & p_1 \downarrow\uparrow\downarrow p_2 \\
 & & & & X_1 & \overset{x_1}{\underset{x_2}{\rightrightarrows}} & X_2 \\
 & & & & & \overset{m}{\longleftarrow} &
\end{array}$$

with contractile pairs as specified in (2.4).

In effect, it suffices to consider a simple transposition of the pair (x_1, x_2) into the category of sets by means of

$(\mathcal{A}(T_1,x_1), \mathcal{A}(T_1,x_2))$ and to apply the remarks made in (2.1) - (2.3). All the mappings defined are «canonical» in the sense that their definition depends only on the universal properties of fiber products and are thus valid in any category satisfying the hypotheses of (2.4).

(2.5) <u>Proposition</u>. Let $\mathcal{A}$ be a category and (x_1,x_2,m): $X_1 \rightleftarrows X_2$ a contractile pair for which the equivalence pairs $\mathcal{R}(x_2m)$, $\mathcal{R}(x_2)$, and $\mathcal{R}(x_1 \hat{\times} x_1)$ (or $\mathrm{Sep}(x_1q_1,x_1q_2)$) of proposition (2.4) exist. Then a co-kernel of the pair (x_1,x_2) exists if, and only if, a co-kernel of the pair (p_1,p_2) exists, in which case they are canonically isomorphic.

In effect let $t: X_2 \longrightarrow T$ be an arrow of $\mathcal{A}$ such that $tp_1 = tp_2$, then $tp_1(\langle x_1,x_2\rangle) = tp_2(\langle x_1,x_2\rangle)$ (where $\langle x_1,x_2\rangle$ is the arrow defined through the relation «$x_2mx_1 = x_2mx_2$») and thus $tx_1 = tx_2$. Conversely, if $t: X_2 \longrightarrow T$ is such that $tx_1 = tx_2$, one has

$$tx_1 = tx_2q_1 = tx_2q_2 = tx_1q_2, \text{ since } x_2q_1 = x_2q_2.$$

Thus $t(p_1x_1 \hat{\times} x_1) = t(p_2x_1 \hat{\times} x_1)$ by definition of $x_1 \hat{\times} x_1$ and so $tp_1(x_1 \hat{\times} x_1) = tp_2(x_1 \hat{\times} x_1)$. But $x_1 \hat{\times} x_1$ is an epimorphism (as a retraction), which allows us to conclude that $tp_1 = tp_2$.

We have thus shown that, for all $T \in \mathcal{O}b(\mathcal{A})$, $\mathrm{Ker}(\mathcal{A}(T,x_1), \mathcal{A}(T,x_2)) = \mathrm{Ker}(\mathcal{A}(T,p_1), \mathcal{A}(T,p_2))$, whence the proposition.

The base of the proof of our «tripleability theorem»

rests on the propositions (2.4) and (2.5). We finish this section, however, with a proposition which we will use later.

(2.6) Proposition. If $X_1 \overset{x_1}{\underset{x_2}{\rightrightarrows}} X_2$ is an equivalence pair in a category $\mathcal{A}$ which is contractile with contraction m: $X_2 \longrightarrow X_1$, the commutative square

$$\begin{array}{ccc} X_1 & \xrightarrow{x_2} & X_2 \\ {\scriptstyle x_1}\downarrow & & \downarrow{\scriptstyle x_2 m} \\ X_2 & \xrightarrow[x_2 m]{} & X_2 \end{array}$$

is cartesian, i.e. $\mathcal{R}(x_2m)$ exists and $\langle x_1,x_2\rangle : X_1 \longrightarrow \mathcal{R}(x_2m)$ is an isomorphism.

As usual, it will suffice to prove the proposition in the category of sets:

In effect, by definition of a contractile pair $x_2mx_1 = x_2mx_2$, so that $\langle x_1,x_2\rangle : X_1 \longrightarrow \mathcal{R}(x_2m)$ is an injection whose image is the graph of an equivalence relation on X_2. For the moment we identify X_1 with this image. (x_1,x_2) is contractile and thus for any $(x,y)\in \mathcal{R}(x_2m)$, $m(x) = (x,pr_2m(x)) \in X_1$ and $m(y) = (y,pr_2m(y)) \in X$. As (x_1,x_2) is an equivalence couple, it is symmetric and hence $(y,pr_2m(y)) \in X_1$ implies $(pr_2m(y),y) \in X_1$. But $pr_2m(y) = pr_2m(x)$ and by the transitivity of X_1 we have that $(x,y) \in X_1$. Hence $\langle x_1,x_2\rangle : X_1 \longrightarrow \mathcal{R}(x_2m)$ is a bijection as desired. Q.E.D.

(2.7) Corollary. Each equivalence pair which is contractile is associated with an arrow. In order that it be an effective equivalence pair it is necessary and sufficient that it admit a co-kernel.

This is an immediate consequence of (2.6) and (1.4).

(2.8) We remark that the «galois dual» of (2.7) is also true.

In order that a retraction f be effective, it is necessary and sufficient that an equivalence pair associated with f exist.

3. A variation on Beck's criterion

The proof of our main result will use the following theorem of J. Beck:

(3.0) Theorem. A functor U: $\mathcal{A} \to \mathcal{B}$ is tripleable if and only if U admits a left-adjoint and verifies the following conditions (Beck's criterion):

(a) If $X_1 \overset{x_1}{\underset{x_2}{\rightrightarrows}} X_2$ is a double arrow in $\mathcal{A}$ such that $(U(x_1), U(x_2))$ is a contractile pair with co-kernel $\nu: U(X_2) \to Q$, then

(B) (x_1, x_2) admits a co-kernel $\nu': X_2 \to Q'$ such that $U(\nu') \xrightarrow{\sim} \nu$.

(b) Moreover, if $\nu'': X_2 \to Q''$ is an arrow in $\mathcal{A}$ such that $\nu'' x_1 = \nu'' x_2$ and $U(\nu'') \xrightarrow{\sim} \nu$, then $\nu'' \xrightarrow{\sim} \nu'$.

The part (a) of condition (B) may be abbreviated as "U <u>creates</u> co-kernels of (U-)-contractile pairs", while the entire condition (B) will be referred to as "U-<u>generates</u> co-kernels of (U-) contractile pairs". Condition (B) is easily seen to be equivalent to the conjunction of condition (a) and the condition

(b') If $X_1 \overset{x_1}{\underset{x_2}{\rightrightarrows}} X_2 \xrightarrow{\nu'} Q'$ is a complex (i.e. $\nu' x_1 = \nu' x_2$) in $\mathcal{a}$ such that $U(X_1) \rightrightarrows U(X_2) \xrightarrow{U(\nu')} U(Q')$ is co-exact (i.e. $U(\nu')$ is a co-kernel of $(U(x_1),U(x_2))$), with $(U(x_1),U(x_2))$ contractile, then $X_1 \rightrightarrows X_2 \rightarrow Q'$ is co-exact.

(b') will be referred to as "U <u>reflects</u> co-kernels of (U-) contractile pairs". Finally we note that condition (a) may also be split into two parts, one assuring the existence of co-kernels of U-contractile pairs, and the other assuring their conservation under U, i.e. the condition

(B') Let $X_1 \overset{x_1}{\underset{x_2}{\rightrightarrows}} X_2$ be a double arrow in $\mathcal{a}$ such that $(U(x_1),U(x_2))$ is contractile and semi-effective (i.e. admits a co-kernel). Then

1° there exists a $\nu' : X_2 \rightarrow Q'$ such that $X_1 \rightrightarrows X_2 \rightarrow Q'$ is co-exact;

2° for any complex $\mathcal{C} : X_1 \overset{x_1}{\underset{x_2}{\rightrightarrows}} X_2 \rightarrow X_3$ in $\mathcal{a}$ involving (x_1,x_2), $\mathcal{C}$ is co-exact $\Leftrightarrow U(\mathcal{C})$ is co-exact.

For the sake of completeness, recall that the original phrasing of Beck was simply condition (a) together with the requirement that U <u>reflected</u> <u>isomorphisms</u>. The proof that all of these conditions are equivalent is, of course, elementary and is left to the reader. We note also that one may obtain the condition for isomorphism in place of simple equivalence by replacing the isomorphism condition on the co-kernels with equality and the same observation will be true for our variant.

(3.1) What we wish to show here is that if the base category admits square fiber products, then the condition (B) may be replaced with the much restricted condition that only requires (B) to be verified for <u>equivalence</u> <u>pairs</u> which are (U-) contractile instead of for <u>arbitrary</u> such pairs. There is a small price to be paid, however: we must guarantee the existence in $\mathcal{A}$ of certain indispensible kernel pairs.

(3.2) <u>Theorem</u>. If U: $\mathcal{A} \to \mathcal{B}$ is a functor into a category $\mathcal{B}$ which admits square fiber products, then U is tripleable if and only if U admits a left-adjoint F and verifies the following conditions:

(a) If $X_1 \underset{x_2}{\overset{x_1}{\rightrightarrows}} X_2$ is an equivalence pair in $\mathcal{A}$ such that $(U(x_1),U(x_2))$ is a contractile pair with co-kernel ν: $U(X_2) \to Q$, then (x_1,x_2) admits a co-kernel ν': $X_2 \to Q'$ such that $U(\nu') \xrightarrow{\sim} \nu$. Moreover,

(B*) if ν'': $X_2 \to Q''$ is an arrow in $\mathcal{A}$ such that $\nu'' x_1 = \nu'' x_2$ and $U(\nu'') \xrightarrow{\sim} \nu$, then $\nu'' \xrightarrow{\sim} \nu'$.

(b) If $X_1 \overset{x_1}{\underset{x_2}{\rightrightarrows}} X_2$ is a double arrow in $\mathcal{A}$ such that $U(x_1,x_2)$ admits a separator, then (x_1,x_2) also admits a separator.

The condition (b) in the presence of the existence of square fiber products in $\mathcal{B}$ has the effect of requiring the existence of square fiber products in $\mathcal{A}$ after the remarks of (1.3). In the presence of a left-adjoint F we have in fact that (b) may be replaced by the statement that U <u>creates</u> separators and in particular square fiber products. As we shall see later it may in fact be replaced by the condition the U <u>generates</u> separators. This is certainly the case for a tripleable functor since it is easy to verify that if U is tripleable, then U <u>generates</u> arbitrary inverse limits and in particular separators of pairs. This last remark together with the fact that (B) implies <u>à fortiori</u> the condition (a) of (B*) completes the trivial part of the demonstration: i.e. if U is tripleable then (B*) is verified. The more interesting part is the converse whose proof we begin with the following

(3.3) <u>Remark</u>. If U: $\mathcal{A} \to \mathcal{B}$ is a functor into a category $\mathcal{B}$ with square fiber products which admits a left-adjoint and verifies the condition (B*) then U verifies the condition

(R): for all $f \in \mathcal{A}r(\mathcal{A})$, U(f) a retraction $\Rightarrow$ f is an effective epimorphism.

In effect, if f: X $\to$ Y is an arrow in $\mathcal{A}$ and U(f) is a retraction then $\mathcal{R}$(U(f)) is a contractile pair which has U(f) as a co-kernel (2.2). By (B*b), $\mathcal{R}(f)$ exists and in the presence of a left-adjoint projects (up to isomorphism) on $\mathcal{R}(U(f))$. Thus the sequence $\mathcal{R}(f) \rightrightarrows X \xrightarrow{f} Y$ projects on a co-exact sequence of the required type and is thus itself co-exact, i.e. f is effective (1.4).

(3.4) <u>Lemma</u>. If $\mathcal{A}$ and $\mathcal{B}$ are categories with square fiber products and U: $\mathcal{A} \to \mathcal{B}$ is a functor with left-adjoint F: $\mathcal{B} \to \mathcal{A}$ and which verifies the condition (R) of (3.3), then

(a) the functor U is faithful;

(b) a double arrow (x_1,x_2): $X_1 \rightrightarrows X_2$ is an equivalence pair if and only if $(U(x_1),U(x_2))$ is such a pair; and

(c) $X_1 \underset{x_2}{\overset{x_1}{\rightrightarrows}} X_2 \xrightarrow{x_3} X_3$ is exact (i.e. is a square fiber product) if and only if the diagram $U(X_1) \rightrightarrows U(X_2) \to U(X_3)$ is exact.

The demonstration of the faithfulness of U is well known. In effect, the canonical mapping of $\mathcal{A}(T,X)$ into $\mathcal{B}(U(T),U(X))$ is that associated with the arrow β_T: $FU(T) \to T$ by the Yoneda Lemma. The image of β_T under the functor U is a retraction and by the hypothesis (R), β_T is an epimorphism, i.e. the canonical mapping β_T^*: $\mathcal{A}(T,X) \to \mathcal{B}(U(T),U(X))$ is an injection for all objects T and X in $\mathcal{A}$.

If $\mathcal{R}(\beta_T) \rightrightarrows FU(T)$ exists, (R) assures us that β_T is an effective epimorphism. By the same token, the canonical arrow

$\beta_{\mathcal{R}(\beta_T)}: FU(\mathcal{R}(\beta_T) \to \mathcal{R}(\beta_T)$ is an epimorphism, thus the diagram of sets

$$\mathcal{A}(T,X) \hookrightarrow \mathcal{A}(FU(T),X) \rightrightarrows \mathcal{A}(FU(\mathcal{R}(\beta_T),X)$$

is exact, whatever be T and X.

Let $X_1 \rightrightarrows X_2$ be an equivalence pair (resp. an equivalence pair associated with an arrow $x_3: X_2 \to X_3$). Then the existence of the left-adjoint implies the same respective facts for the image pair $(U(x_1),U(x_2))$.

Conversely, if such is the case for $(U(x_1),U(x_2))$, we will have the exact diagram

$$\mathcal{B}(Y,U(X_1)) \hookrightarrow \mathcal{B}(Y,U(X_2)) \times \mathcal{B}(Y,U(X_2)) \rightrightarrows \mathcal{B}(Y,U(X_3))$$

for any $Y \in \mathcal{O}b(\mathcal{B})$ in the case for (c) and, in the case (b), simply the fact that $\mathcal{B}(Y,U(X_1))$ has as its image the graph of an equivalence relation on $\mathcal{B}(Y,U(X_2))$. In particular this will be true where Y is U(T) and also where Y is $U(\mathcal{R}(\beta_T))$.

Using the adjunction bijection, we have for the case (c) a diagram of bi-simplicial type.

$$\begin{array}{ccccc}
\mathcal{A}(T,X_1) & \longrightarrow & \mathcal{A}(T,X_2) \times \mathcal{A}(T,X_2) & \rightrightarrows & \mathcal{A}(T,X_3) \\
\downarrow & \text{(D)} & \downarrow & & \downarrow \\
\mathcal{A}(FU(T),X_1) & \hookrightarrow & \mathcal{A}(FU(T),X_2) \times \mathcal{A}(FU(T),X_2) & \rightrightarrows & \mathcal{A}(FU(T),X_3) \\
\downarrow\downarrow & & \downarrow\downarrow & & \downarrow\downarrow \\
\mathcal{A}(FU(\mathcal{R}(\beta_T)),X_1) & \hookrightarrow & \mathcal{A}(FU(\mathcal{R}(\beta_T),X_2) \times \mathcal{A}(FU(\mathcal{R}(\beta_T) & \rightrightarrows & \mathcal{A}(FU(\mathcal{R}\beta_T),X_3)
\end{array}$$

in which all of the columns and the lines with FU($\mathcal{R}$(pr)) and FU(T) are exact, and in the case (b) the same diagram but with the column involving X_3 suppressed. But by Lemma (0.3) the diagram (D) is cartesian which makes $\mathcal{A}(T,X_1)$ bijectively equivalent to the inverse image of an equivalence relation and thus itself bijectively equivalent to an equivalence relation (which completes the proof of (b)), and by the corollary (0.4) we have that the line involving T is exact which completes the proof of (c). Q.E.D.

We note in passing that the proof of part (c) may be modified in an obvious fashion to prove that a projective cone in $\mathcal{A}$ is an inverse limit of a diagram $\mathcal{D}$ if and only if its image under U is an inverse limit of the diagram U($\mathcal{D}$).

(3.5) Proof of Theorem (3.2) (contd.). We now give the proof that (B*) is sufficient for the creation of co-kernels of U-contractile pairs. The basic idea will be to apply the proposition (2.4) in the base category $\mathcal{B}$ and use the condition (B*) to recreate as much of it as is necessary in $\mathcal{A}$.

In effect, let (x_1,x_2) be a double arrow in $\mathcal{A}$ such that $(U(x_1),U(x_2))$ is contractile with co-kernel $\nu: U(X_2) \longrightarrow Q$. We show that there exists an arrow $\nu': X_2 \longrightarrow Q'$ in $\mathcal{A}$ which is a co-kernel of (x_1,x_2) and which projects up to isomorphism onto ν:

First form the pair $\mathcal{R}(x_2) \overset{q_1}{\underset{q_2}{\rightrightarrows}} X_1$ which exists in $\mathcal{A}$ by (B*(b)) and consider the pair $(x_1q_1,x_1q_2): \mathcal{R}(x_2) \rightrightarrows X_2$. A separator of $(U(x_1q_1),U(x_1q_2))$ exists as the equivalence pair associated with $U(x_1) \hat{\rightleftarrows} U(x_1)$. Thus, again by (B*(b)) a separator (r_1,r_2) of (x_1q_1,x_1q_2) exists in $\mathcal{A}$ and projects up to isomorphism onto the contractile pair $\mathcal{R}(U(x_1) \hat{\rightleftarrows} U(x_1))$. Now this separator is an equivalence pair; thus by (B*(a)) it admits a co-kernel $\hat{x}_1$: $\mathcal{R}(x_2) \longrightarrow R$ which projects up to isomorphism on the retraction $U(x) \hat{\rightleftarrows} U(x_1)$: $\mathcal{R}(U(x_2)) \longrightarrow \mathcal{R}(U(x_2)m)$. Applying the fact that (r_1,r_2) is a separator (so that $x_1q_1r_1 = x_1q_1r_2$ and $x_1q_2r_1 = x_1q_2r_2$) and the fact that $\hat{x}_1$ is a co-kernel of this separator, we have the existence of a pair of arrows (p_1,p_2): $R \rightrightarrows X_2$ such that $p_1\hat{x}_1 = x_1q_1$ and $p_2\hat{x}_1 = x_1q_2$. But $U(R) \xrightarrow{\sim} \mathcal{R}(U(x_2)m)$, thus the pair (p_1,p_2) has for image a contractile equivalence pair; it is thus itself an equivalence pair (by Lemma (3.4b)) which has for image a contractile pair with co-kernel ν: $U(X_2) \longrightarrow Q$. It thus admits a co-kernel ν': $X_2 \longrightarrow Q'$ such that $U(\nu') \longrightarrow \nu$. But now U is faithful again, by Lemma (3.4a) and thus $U(\nu' x_1) = U(\nu')U(x_1) = U(\nu')U(x_2) = U(\nu' x_2)$ implies that $\nu' x_1 = \nu' x_2$. But given any f: $X_2 \longrightarrow T$ such that $fx_1 = fx_2$, on has equally that $fp_1 = fp_2$, since (p_1,p_2) is effective. This completes the proof of creation. The remainder of the proof is trivial since the faithfulness is sufficient to guarantee that $\mathrm{Ker}(\mathcal{A}(x_1,T), \mathcal{A}(x_2,T)) = \mathrm{Ker}(\mathcal{A}(p_1,T), \mathcal{A}(p_2,T))$ for all $T \in \mathcal{O}b(\mathcal{A})$ and thus the remaining property is implied by the second part of (B*(a)). Q.E.D

4. Tripleability over "set-like" categories

If the base category B has certain "set-like" properties, the preceding theorem may be modified in a fashion which symmetrizes the condition (B*). For example we have the following proposition whose proof makes heavy use of Lemma (3.4).

(4.1) Proposition. Let $\mathcal{A}$ and $\mathcal{B}$ be categories with square fiber products and U: $\mathcal{A} \to \mathcal{B}$ a functor with a left-adjoint. If the category $\mathcal{B}$ is such that every equivalence pair in B is contractile and semi-effective (and thus effective by corollary (2.7)), the following statements are equivalent:

1° For any equivalence pair $R \rightrightarrows S$, and any arrow $S \to T$ such that $R \rightrightarrows S \to T$ is a complex

(a) there exists an arrow ν: $S \to Q$ such that $R \rightrightarrows S \to Q$ is co-exact (i.e. $R \rightrightarrows S$ is semi-effective)

(b) $R \rightrightarrows S \to T$ is co-exact $\Longleftrightarrow U(R) \rightrightarrows U(S) \to U(T)$ is co-exact;

2° for any equivalence pair $R \rightrightarrows S$ and any $S \to T$ such that $R \rightrightarrows S \to T$ is a complex

(a) $R \rightrightarrows S$ is effective, and

(b) $R \rightrightarrows S \to T$ is bi-exact $\Longleftrightarrow U(R) \rightrightarrows U(S) \to U(T)$ is bi-exact;

3° for any equivalence pair $R \rightrightarrows S$ and any arrow f in $\mathcal{A}$,

(a) $R \rightrightarrows S$ is effective, and

(b) f is effective $\Longleftrightarrow$ U(f) is effective ($\Longleftrightarrow$ U(f) is a retraction);

4° for any pair $R \rightrightarrows S$ and any arrow f in $\mathcal{A}$,

(a) $R \rightrightarrows S$ is effective $\Longleftrightarrow U(R) \rightrightarrows U(S)$ is effective, and

(b) f is effective $\longleftrightarrow$ U(f) is effective ($\longleftrightarrow$ U(f) is a retraction)

5° U verifies the condition B*(a) of theorem (3.2).

(1° $\Rightarrow$ 2°). Let f: A $\longrightarrow$ B be an arrow such that U(f) is a retraction. Since $\mathcal{A}$ admits square fiber products, a pair $\mathcal{R}$(f) $\rightrightarrows$ A exists such that the sequence U($\mathcal{R}$(f)) $\rightrightarrows$ U(A) $\longrightarrow$ U(f) is exact. U(f) is a retraction, hence U($\mathcal{R}$(f)) $\rightrightarrows$ U(A) is contractile and U($\mathcal{R}$(f)) $\rightrightarrows$ U(A) $\longrightarrow$ U(f) is also co-exact (2.8). By 1b we have that $\mathcal{R}$(f) $\rightrightarrows$ A $\xrightarrow{f}$ B is co-exact and hence f is effective. Thus Lemma (3.4) is applicable. Let R $\rightrightarrows$ S be an equivalence pair; by 1°a, R $\rightrightarrows$ S admits a co-kernel ν: S $\longrightarrow$ Q and by 1°b U(R) $\rightrightarrows$ U(S) $\longrightarrow$ U(Q) is co-exact. But U has a left-adjoint, hence U(R) $\rightrightarrows$ U(S) is an equivalence pair which is thus by hypothesis contractile. By corollary (2.7) U(R) $\rightrightarrows$ U(S) $\longrightarrow$ U(Q) is exact. By Lemma 3.4, R $\rightrightarrows$ S $\longrightarrow$ Q is also exact and R $\rightrightarrows$ S is effective. Suppose that a complex R $\rightrightarrows$ S $\longrightarrow$ Q is also exact and R $\rightrightarrows$ S $\longrightarrow$ T is bi-exact. By 1°b the complex U(R) $\rightrightarrows$ U(S) $\longrightarrow$ U(T) is co-exact and since U has a co-adjoint it is also exact and thus bi-exact. Similarly if U(R) $\rightrightarrows$ U(S) $\longrightarrow$ U(T) is bi-exact, 1°b guarantees that R $\rightrightarrows$ S $\longrightarrow$ T is co-exact and Lemma (3.4) will give that it is also exact.

(2° $\Rightarrow$ 3°). 2°(a) $\Rightarrow$ 3°(a) is a tautology. Suppose that f: A $\longrightarrow$ B is effective, then $\mathcal{R}$(f) $\rightrightarrows$ A $\xrightarrow{f}$ B is bi-exact, and by 2°(b) we have that U($\mathcal{R}$(f) $\rightrightarrows$ U(A) $\longrightarrow$ U(B) is bi-exact, hence U(f) is effective. If U(f) is effective, then $\mathcal{R}$(U(f)) $\rightrightarrows$ U(A) $\longrightarrow$ U(B) is bi-exact. But $\mathcal{A}$ has square fiber products and U has a left-adjoint; hence $\mathcal{R}(f) \rightrightarrows A \xrightarrow{f} B$ is a complex which projects onto a bi-exact sequence. It is thus by 2°(b) itself bi-exact and f is effective.

(3° $\Rightarrow$ 4°). Note that by (2.2) 3°(b) $\Rightarrow$ 4° is a tautology. Suppose that $R \rightrightarrows S$ is effective, then $R \rightrightarrows S \xrightarrow{\nu} \mathrm{Coker}(R \rightrightarrows S)$ is bi-exact and ν is effective; but then 3°(b) $U(\nu)$ is effective and so is $U(R) \rightrightarrows U(S)$ since $U(R) \rightrightarrows U(S) \longrightarrow (\mathrm{Cok}(R \rightrightarrows S)$ is exact (because of the existence of the left-adjoint to U). Conversely, if $U(R) \rightrightarrows U(S)$ is effective, $U(R) \rightrightarrows U(S) \xrightarrow{\nu} \mathrm{Cok}(U(R) \rightrightarrows U(S))$ is bi-exact. By (2.2) any effective epimorphism is necessarily a retraction since any equivalence pair in $\mathcal{B}$ is contractile, in particular that associated with the effective epimorphism. Consequently Lemma (3.4) is directly applicable and hence $R \rightrightarrows S$ is an equivalence pair. By 3°(a) $R \rightrightarrows S$ is effective.

(3° $\Rightarrow$ 4°). 3°(b) $\Rightarrow$ 4°(b) is a tautology. By (2.2) any effective epimorphism in $\mathcal{B}$ is a retraction since any equivalence pair is contractile, in particular that associated with the effective epimorphism. Lemma (3.4) is thus directly applicable. Suppose then that $U(R) \rightrightarrows U(S)$ is effective. It is thus an equivalence pair and by (3.4) so is $R \rightrightarrows S$. But any equivalence pair in $\mathcal{A}$ is effective by 3°(a). Conversely if $R \rightrightarrows S$ is effective it is an equivalence pair and equivalence pairs are conserved in the presence of a left-adjoint. But any equivalence pair in $\mathcal{B}$ is effective by hypothesis, in particular $U(R) \rightrightarrows U(S)$.

($4° \Rightarrow 5°$). Let $R \rightrightarrows S$ be an equivalence pair in $\mathcal{A}$ which projects on a U-contractile semi-effective pair in $\mathcal{B}$. Since U has a left-adjoint, $U(R) \rightrightarrows U(S)$ is an equivalence pair and is hence effective by hypothesis. By 4°(a) $R \rightrightarrows S$ admits a co-kernel and is in fact effective; thus $R \rightrightarrows S \xrightarrow{\nu} \mathrm{Cok}(R \rightrightarrows S)$ is bi-exact and ν is an effective epimorphism; by 4°(b) $U(\nu)$ is an effective epimorphism and the complex $U(R) \rightrightarrows U(S) \xrightarrow{U(\nu)} U(\mathrm{Cok}(R \rightrightarrows S))$ is bi-exact since U has a left-adjoint; thus $U(\nu) \xrightarrow{\sim} \mathrm{Cok}(U(R) \rightrightarrows U(S)$ as desired. Let $R \rightrightarrows S \xrightarrow{\nu'} T$ be a complex such that $U(\nu')$ is a co-kernel of $U(R) \rightrightarrows U(S)$. Since $U(R) \rightrightarrows U(S)$ is contractile by corollary 2.7, the sequence $U(R) \rightrightarrows U(S) \longrightarrow U(T)$ is bi-exact. Now since 4°(b) holds, Lemma 3.4 is applicable and $R \rightrightarrows S \longrightarrow T$ is exact. But 4°(b) says that ν is itself effective since $U(\nu)$ is effective; thus $\nu : S \longrightarrow T$ is isomorphic to $\mathrm{Cok}(R \rightrightarrows S)$.

($5° \Rightarrow 1°$). Let $R \rightrightarrows S$ be an equivalence pair in $\mathcal{A}$, $U(R) \rightrightarrows U(S)$ is an equivalence pair in $\mathcal{B}$ which is effective and contractile; thus $R \rightrightarrows S$ admits a co-kernel, i.e. 1°(a) holds. Let $R \rightrightarrows S \longrightarrow T$ be co-exact. Since $R \rightrightarrows S$ is an equivalence pair $U(R) \rightrightarrows U(S)$ is also an equivalence pair which is thus semi-effective by the hypothesis on $\mathcal{B}$. By 5° $R \rightrightarrows S$ admits a co-kernel ν such that $U(\nu)$ is isomorphic to the given co-ker of $U(R) \rightrightarrows U(S)$. But $\nu \xrightarrow{\sim} (S \longrightarrow T)$ as co-kernels and thus $U(R) \rightrightarrows U(S) \longrightarrow U(T)$ is co-exact. Suppose that $U(R) \rightrightarrows U(S) \longrightarrow U(T)$ is co-exact, then by 5° again $R \rightrightarrows S \longrightarrow T$ is co-exact since $U(R) \rightrightarrows U(S)$ is always contractile as an equivalence pair. Q.E.D.

(4.2) Corollary. Let U: $\mathcal{A} \longrightarrow \mathcal{B}$ be a functor into a category $\mathcal{B}$ which has separators of double arrows and in which every equivalence pair is contractile and admits a co-kernel (and is thus effective). Under this condition, the functor U is tripleable if and only if $\mathcal{A}$ admits separators of pairs, U admits a left-adjoint, and any one of the equivalent conditions 1°-5° of Proposition (4.1) is verified.

This last corollary has as its own corollary the characterization theorem of Linton.

(4.4) Theorem (Linton (1965)). Let U: $\mathcal{A} \longrightarrow$ (ENS) be a functor. U is tripleable if and only if U has a left-adjoint and the following three conditions are satisfied:

1° $\mathcal{A}$ has square fiber products and co-kernels;

2° $R \rightrightarrows S$ is a kernel pair $\Longleftrightarrow U(R) \rightrightarrows U(S)$ is a kernel pair

3° f: $A \longrightarrow B$ is a co-kernel $\Longleftrightarrow U(f)$ is a co-kernel

where $R \rightrightarrows S$ is a kernel pair means there exists an arrow f: $S \longrightarrow T$ such that $R \rightrightarrows S \longrightarrow T$ is exact and f: $S \longrightarrow T$ is a co-kernel means there exists a pair $R \rightrightarrows S$ such that $R \rightrightarrows S \longrightarrow T$ is co-exact. This is immediate since if a category has co-kernels, $R \rightrightarrows S$ a kernel pair $\Longleftrightarrow R \rightrightarrows S$ is effective and if it has square fiber products f: $A \longrightarrow B$ is a coker $\Longleftrightarrow$ f is effective. Since the existence of co-kernels and square fiber products implies that of separators (1.3), the theorem is immediate (4.1 condition 4°).

For the necessity of the existence of co-kernels see (7.) below.

5. Categories tripleable over functor categories

In this section we use techniques derived from the theory of triples to characterize those categories which are tripleable over categories of the form $\hat{\mathcal{C}} = \mathrm{CAT}(\mathcal{C}^{op},(\mathrm{ENS}))$, where $\mathcal{C}$ is some ($\mathcal{U}$-) small category. The entire section may be viewed as a preliminary study in the lines of Gabriel's (1966) treatment which suggested its possible interest.

(5.0) Let $\mathcal{C}$ be a ($\mathcal{V}$-) small category. Any functor $U: \mathcal{X} \to \hat{\mathcal{C}}$ (= $\mathrm{CAT}(\mathcal{C}^{op},(\mathrm{ENS}))$) which has a left-adjoint $F: \hat{\mathcal{C}} \to \mathcal{X}$ is isomorphic to the functor $\widehat{Fh_{\mathcal{C}}}\, h_{\mathcal{X}}$ induced by restriction of $Fh_{\mathcal{C}}: \hat{\mathcal{X}} \to \hat{\mathcal{C}}$ by the Yoneda embedding $h_{\mathcal{X}}: \mathcal{X} \to \hat{\mathcal{X}}$, since for any objects X in $\mathcal{X}$ and c in $\mathcal{C}$, one has the functorial isomorphisms

$$\mathcal{X}(F(h_c),X) \xrightarrow{\sim} \mathcal{C}(h_c,U(X)) \xrightarrow{\sim} U(X)(c).$$

We may thus restrict our attention to functors $S: \mathcal{C} \to \mathcal{X}$ and their associates $\langle\!\langle X \mapsto \mathcal{X}(S(c),X)\rangle\!\rangle$ induced by S from $\mathcal{X}$ into $\hat{\mathcal{C}}$ and make the following

(5.1) <u>Definition</u>. A functor $S: \mathcal{C} \to \mathcal{X}$ will be called a <u>co-frame</u> (of $\mathcal{C}$) provided any small diagram $\Theta: \mathcal{D} \to \mathcal{C}$ is such that $\varinjlim S\Theta$ exists in $\mathcal{X}$. (i.e. any small diagram in $\mathcal{C}$ has a co-limit in $\mathcal{X}$).

The interest of this definition for us here is the following.

(5.2) **Proposition.** For any small category $\mathcal{C}$ and any functor $S: \mathcal{C} \to \mathcal{X}$ the following statements are equivalent:

1° $S: \mathcal{C} \to \mathcal{X}$ is a co-frame;

2° the functor $\hat{S}h_{\mathcal{X}}: \mathcal{X} \to \hat{\mathcal{C}}$ $(= \mathrm{CAT}(\mathcal{C}^{op}, (\mathrm{ENS})))$ admits a left-adjoint

In effect, for any $F \in \mathcal{O}b(\hat{\mathcal{C}})$ any $X \in \mathcal{O}b(\mathcal{X})$, the sequence

$$\hat{\mathcal{C}}(F, h_X S) \hookrightarrow \prod_{c \in \mathcal{O}b(\mathcal{C})} \mathrm{ENS}(F(c), \mathcal{X}(S(c), X)) \overset{\ell_1}{\underset{\ell_2}{\rightrightarrows}} \prod_{\theta \in \mathcal{A}r(\mathcal{C})} \mathrm{ENS}(F\mathcal{B}(\theta), \mathcal{X}(S(\mathcal{S}(\theta), X))$$

where ℓ_1 is defined by $(\mathrm{pr}_{\mathcal{S}(\theta)}(\mathcal{Z})F(\theta))_{\theta \in \mathcal{A}r(\mathcal{C})}$ and ℓ_2 by $(h_X S(\theta) \mathrm{pr}_{\mathcal{B}(\theta)}(\mathcal{Z}))_{\theta \in \mathcal{A}r(\mathcal{C})}$ (with $\mathcal{S}, \mathcal{B}: \mathcal{A}r(\mathcal{C}) \rightrightarrows \mathcal{O}b(\mathcal{C})$ denoting, respectively, the domain and co-domain functions of $\mathcal{C}$) is exact, but then the equivalent sequence

$$\hat{\mathcal{C}}(F, h_X S) \hookrightarrow \prod_{c \in \mathcal{O}b(c)} \mathcal{X}(S(c), X)^{F(c)} \rightrightarrows \prod_{\theta \in \mathcal{A}r(\mathcal{C})} \mathcal{X}(S(\mathcal{S}(\theta), X)^{F(\mathcal{B}(\theta))}$$

is exact.

Now again by definition of natural transformation, the sequence

$$\mathrm{Nat}(S\mathcal{S}_F^{\#}, \mathcal{X}(X)) \hookrightarrow \prod_{(c,r) \in \mathcal{O}b(\mathcal{C}/F)} \mathcal{X}(S\mathcal{S}_F^{\#}(c,r), X) \rightrightarrows \prod_{\xi \in \mathcal{A}r(\mathcal{C}/F)} \mathcal{X}(S\mathcal{S}_F^{\#}(\mathcal{S}(\xi)), X)$$

is exact where $S_F^{\#}: \mathcal{C}/F \to \mathcal{C}$ is the small diagram in $\mathcal{C}$ defined by the cartesian square

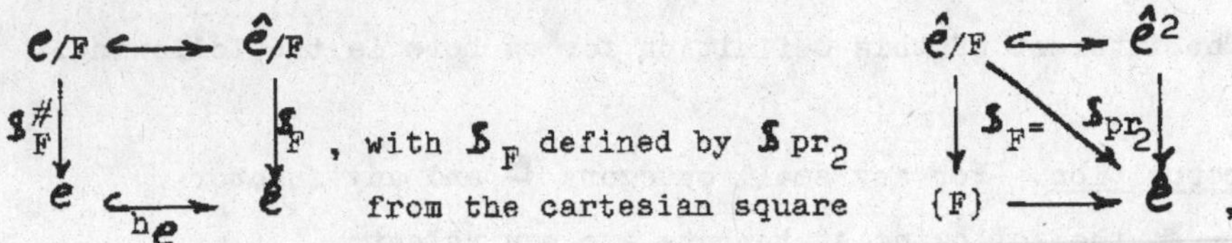

and $\mathcal{K}(X)$: $\mathcal{C}/F \to \mathfrak{X}$ is just the constant diagram for the object X in $\mathfrak{X}$. The last sequence is the defining sequence for $\varinjlim S\mathfrak{S}^{\#}_F$ in $\mathfrak{X}$ and as the category $\mathcal{C}/F$ may be identified with that category whose objects are those pairs (c,γ) with $c \in \mathcal{O}b(\mathcal{C})$ and $\gamma \in F(c)$, and whose arrows are those arrows θ: $\mathcal{B} \to C$ in C such that $F(\theta)(\beta) = \gamma$ we see that we have a canonical bijection

$$\mathrm{Nat}(S\mathfrak{S}^{\#}_F, \mathcal{K}(X)) \xrightarrow{\sim} \hat{\mathcal{C}}(F, h_X S)$$

so that a left-adjoint of $\hat{S}h_{\mathfrak{X}}$ exists if and only if the co-limit of the small diagram $S\mathfrak{S}^{\#}_F$: $\mathcal{C}/F \to \mathfrak{X}$ exists in $\mathfrak{X}$ for all $F \in \hat{\mathcal{C}}$. Thus 1° $\Rightarrow$ 2°. Conversely, suppose that $\hat{S}h_{\mathfrak{X}}$ has a left-adjoint F, then as the diagram

is commutative up to isomorphism (5.0) and any small diagram $\mathcal{D}$: $\mathcal{D} \to \mathcal{C}$ has a co-limit in $\hat{\mathcal{C}}$, we have that $F \varinjlim h_{\mathcal{C}}\mathcal{D} \xrightarrow{\sim} \varinjlim Fh_{\mathcal{C}}\mathcal{D} \xrightarrow{\sim} \varinjlim S\mathcal{D}$, and the implication 2° $\Rightarrow$ 1° is established.

Proposition (5.2) implies, in particular, that any small family of objects of $\mathfrak{X}$ each of which is of the form S(c) for some $c \in \mathcal{O}b(\mathcal{C})$ admits a co-product, and that certain co-kernels of double arrows between such co-products exist.

REMARK: It is convenient to have the explicit formulae used in the preceding proof, but the theorem can be established more quickly:

Since $h_{\mathcal{C}}$ is fully faithful, any functor $K \in \hat{\mathcal{C}}$ is such that

$$K \xrightarrow{\sim} \varinjlim \mathcal{S}_K \xrightarrow{\sim} \varinjlim \mathcal{S}_K h^{\#}_{\mathcal{C}} \xrightarrow{\sim} \varinjlim h_{\mathcal{C}} \mathcal{S}^{\#}_K ,$$

so that if $Sh_{\mathfrak{X}}$ has a left-adjoint F, $F(K) \xrightarrow{\sim} F(\varinjlim h_{\mathcal{C}} S^{\#}_K) \xrightarrow{\sim} \varinjlim Fh_{\mathcal{C}} S^{\#}_K \xrightarrow{\sim} \varinjlim S\mathcal{S}^{\#}_K$. Conversely, if $\varinjlim S\mathcal{S}^{\#}_K$ exists,

$$\mathfrak{X}(\varinjlim S\mathcal{S}^{\#}_K, X) = h_X(\varinjlim S\mathcal{S}^{\#}_K) \xrightarrow{\sim} \varprojlim h_X S\mathcal{S}^{\#}_K \xrightarrow{\sim} \varprojlim_{(c,\gamma)\in\mathcal{C}/K} h_X S(\mathcal{S}^{\#}_K(c,\gamma)) \xrightarrow{\sim}$$

$$\varprojlim_{(c,\gamma)\in\mathcal{C}/K} \hat{\mathcal{C}}(h_{\mathcal{S}^{\#}_K}(c,\gamma), h_X S) \xrightarrow{\sim} \varprojlim_{(c,\gamma)\in\mathcal{C}/K} \hat{\mathcal{C}}(h_{\mathcal{C}}\mathcal{S}^{\#}_K(c,\gamma), h_X S) \xrightarrow{\sim} \hat{\mathcal{C}}(\varinjlim h_{\mathcal{C}}\mathcal{S}^{\#}_K, h_X S)$$

$$\xrightarrow{\sim} \hat{\mathcal{C}}(K, h_X S)$$

as desired.

(5.3) Lemma. If $U: \mathcal{A} \longrightarrow \mathcal{B}$ is a functor with left-adjoint $F: \mathcal{B} \longrightarrow \mathcal{A}$, then for any pair $R \underset{y}{\overset{x}{\rightrightarrows}} S$ in $\mathcal{A}$,

$R \rightrightarrows S$ is U-contractile $\Longleftrightarrow FU(R) \rightrightarrows FU(S)$ is contractile, and

$R \rightrightarrows S \longrightarrow T$ is a U-contractile system $\Longleftrightarrow FU(R) \rightrightarrows FU(S) \longrightarrow FU(T)$ is a contractile system.

If $U(R) \rightrightarrows U(S)$ is contractile with contraction $m: U(S) \longrightarrow U(R)$, then clearly the $F(m): FU(S) \longrightarrow FU(R)$ defines a contraction in $\mathcal{A}$.

Conversely if m: $FU(S) \to FU(R)$ is a contraction in $\mathcal{A}$, one has the sequentially commutative diagram

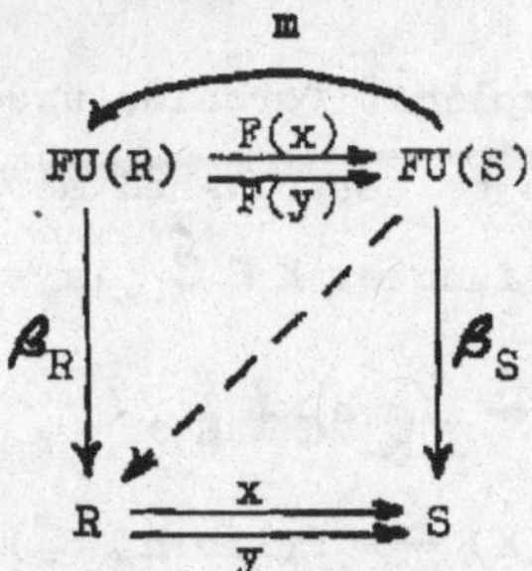

in $\mathcal{A}$ (where β_R and β_S) are the adjunction arrows). If m is a contraction, then one has the chain of equalities

$$y\beta_R m(FU(x) = \beta_S FU(y)mFU(x) = \beta_S FU(y)mFU(y) = y\beta_R mFU(y)$$

as well as

$$x\beta_R m = \beta_S FU(x)m = \beta_S id_{FU(S)} = \beta_S$$

As the arrow $\beta_R m$: $FU(S) \to R$ is equivalent to an arrow $\alpha(\beta_R m)$: $U(S) \to U(R)$, it is easily seen that the equalities imply that $\alpha(\beta_R m)$ is a contraction for $(U(x),U(y))$. Similar remarks hold for the second part since «$X \rightrightarrows Y \to Z$ is a contractile system» is equivalent to «$X \rightrightarrows Y$ is a contractile pair and $X \rightrightarrows Y \to Z$ is co-exact» .

Ordinarily, one is interested in tripleability criteria for a functor U: $\mathcal{A} \to \mathcal{B}$, which do not involve anything more than the existence of a left-adjoint for U. In the application which will follow, however, it is convenient to make explicit reference to this left-adjoint. For this purpose lemma (5.3) gives the following corollary to theorem (3.2).

(5.4) Corollary. Let $U: \mathcal{A} \to \mathcal{B}$ be a functor with left-adjoint F into a category with separators in which every contractile equivalence pair is effective. Under these conditions U is tripleable if and only if the following conditions are satisfied.

1° $\mathcal{A}$ has separators of pairs;

2° Every FU-contractile equivalence pair is effective;

3° For any complex $R \rightrightarrows S \to T$ in $\mathcal{A}$ which has $R \rightrightarrows S$ as a FU-contractile equivalence pair,

$$R \rightrightarrows S \to T \text{ co-exact} \Leftrightarrow FU(R) \rightrightarrows FU(S) \to FU(T) \text{ is co-exact.}$$

(5.5) Definition. Let $S: \mathcal{C} \to \mathcal{X}$ be a functor. A pair $R \rightrightarrows S$ in $\mathcal{X}$ will be called S-contractile provided the co-limits of the functors S_R and S_X defined by the cartesian squares

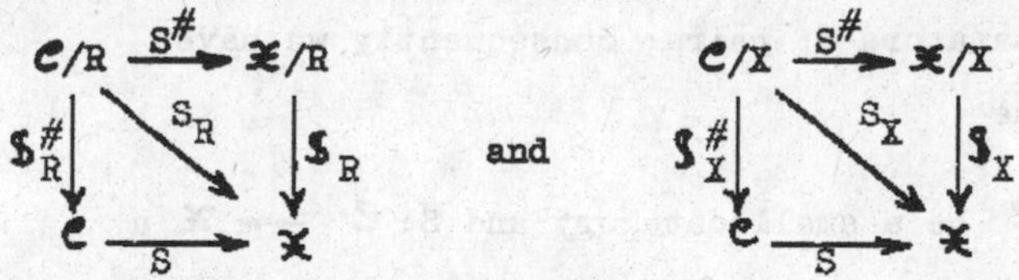

exist and the deduced pair in $\mathcal{X}$

$$\varinjlim S_R \rightrightarrows \varinjlim S_X$$

is contractile.

(5.6) Corollary. Let $\mathcal{C}$ be a small category and $S: \mathcal{C} \to \mathfrak{X}$ a co-frame (5.1). The following statements are equivalent for any pair $R \rightrightarrows X$ in $\mathfrak{X}$:

1° $R \rightrightarrows X$ is $Sh_{\mathfrak{X}}$ - contractile;

2° $R \rightrightarrows X$ is S-contractile.

In effect, it is trivial to verify that the Yoneda-Grothendieck lemma gives an isomorphism of the categories $\mathcal{C}/Z$ and $\mathcal{C}/h_Z S$ as categories over $\mathfrak{X}$ by composition of the projection functors with S. But then $\varinjlim_{\mathcal{C}/Z} S_X \xrightarrow{\sim} \varinjlim_{\mathcal{C}/h_Z S} S\, pr_1 \xrightarrow{\sim} F(h_Z S)$, where $F: \mathcal{C} \to \mathfrak{X}$ is a left-adjoint of $Sh_{\mathfrak{X}}$ (which exists by Proposition (5.2)). The result is thus a corollary of the preceding lemma.

(5.7) Definition. Let $\mathcal{C}$ be a small category and $S: \mathcal{C} \to \mathfrak{X}$ a functor. $\mathfrak{X}$ will be said to be S-tripleable provided the functor $Sh_{\mathfrak{X}}: \mathfrak{X} \to \hat{\mathcal{C}}$ is tripleable.

In $\hat{\mathcal{C}}$ every double arrow is semi-effective (i.e. admits a co-kernel) and $\hat{\mathcal{C}}$ has separators of pairs; consequently we have immediately (from 5.4) the

(5.8) Proposition. Let $\mathcal{C}$ be a small category and $S: \mathcal{C} \to \mathfrak{X}$ a functor. In order that $\mathfrak{X}$ be S-tripleable it is necessary and sufficient that $\mathfrak{X}$ verify the following conditions:

1° S is a co-frame (5.1);

2° $\mathfrak{X}$ has separators of pairs;

3° For any S-contractile equivalence pair $R \rightrightarrows S$ in $\mathfrak{X}$,

(a) $R \rightrightarrows S$ is effective, and

(b) $R \rightrightarrows S \rightarrow T$ is co-exact $\Leftrightarrow \varinjlim S_R \rightrightarrows \varinjlim S_S \rightarrow \varinjlim S_T$ is co-exact.

(5.9) Remark. The last condition of course may be replaced by «for any S-contractile equivalence pair $R \rightrightarrows X$, $R \rightrightarrows X \rightarrow Y$ is bi-exact $\Leftrightarrow$ for all $c \in \mathcal{O}b(\mathcal{C})$ $\mathfrak{X}(S(c),R) \rightrightarrows \mathfrak{X}(S(c),X) \rightarrow \mathfrak{X}(S(c),Y)$ is bi-exact».

(5.10) Lemma. Let $\mathcal{C}$ be a small category and $S: \mathcal{C} \rightarrow \mathfrak{X}$ a functor. Let $\Gamma(S\langle\mathcal{C}\rangle)$ denote the subcategory of $\mathfrak{X}$ generated by the image of $\mathcal{C}$ under S and

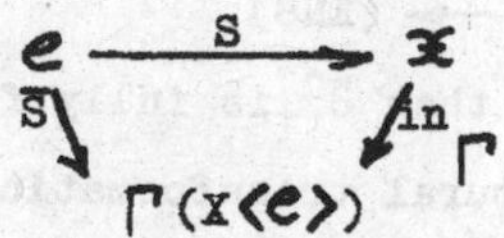

the resulting factorization with in_Γ the canonical inclusion. With this notation, $\mathfrak{X}$ is S-tripleable if and only if $\mathfrak{X}$ is in_Γ-tripleable.

In effect the triangle gives rise to the commutative triangle

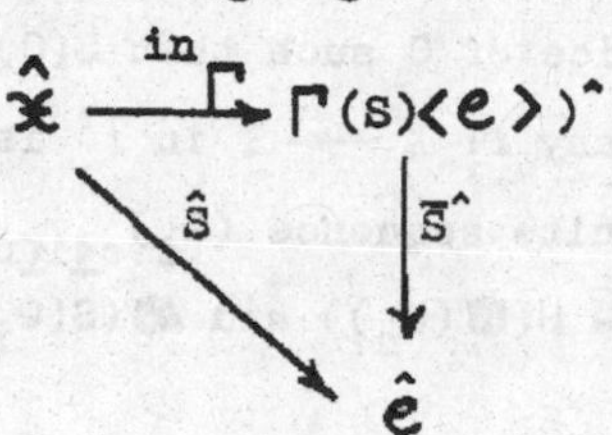

and by composition with $h_{\mathcal{X}}$ to

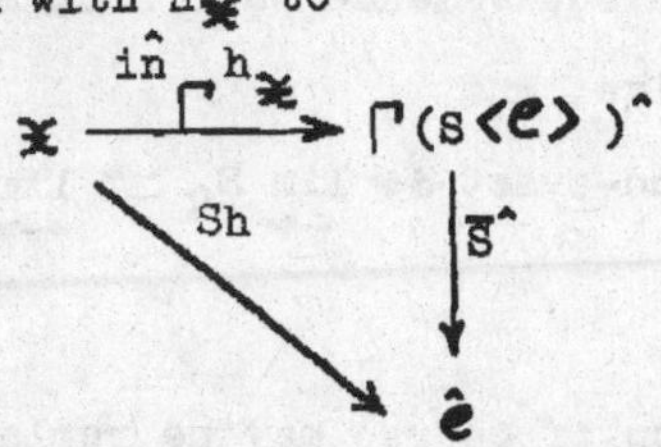

We claim that $\overline{S}^{\wedge}$ is an <u>adjoint</u> <u>section</u> (i.e. $\overline{S}^{\wedge}$ admits a left-adjoint T such that with the adjunction arrow $T\overline{S}^{\wedge} \xrightarrow{\sim} id_{\Gamma(X\langle\mathcal{C}\rangle)^{\wedge}}$). In effect it certainly admits a left-adjoint which may be defined by the Kan formula

$$T(G)(X) = \varinjlim_{\mathcal{C}^{op}/X} G\mathcal{S}^{\#}_{X}$$

for $X \in \mathcal{O}b(\Gamma(S\langle\mathcal{C}\rangle))$ and $G: \mathcal{C}^{op} \longrightarrow (ENS)$.

It will thus suffice to show that $\overline{S}^{\wedge}$ is fully faithful.

If $\psi, \varphi : F_1 \longrightarrow F_2$ are a natural transformations in $\Gamma^{\wedge}(=\Gamma(S\langle\mathcal{C}\rangle)^{\wedge})$ such that $\psi\overline{S}^{op} = \varphi\overline{S}^{op}$, then for all $C \in \mathcal{O}b(\mathcal{C})$, $\psi(\overline{S}(G)) = \varphi(\overline{S}(C)) \longleftrightarrow \psi(S(C)) = \varphi(S(C))$. But every object in Γ has the form $S(C)$ for some $C \in \mathcal{O}b(\mathcal{C})$, hence $\varphi = 4$. Suppose now we are given a natural transformation $\beta: G_1\overline{S}^{op} \longrightarrow G_2\overline{S}^{op}$ in $\hat{\mathcal{C}}$ with G_1 and G_2 in $\hat{\Gamma}$, then define a transformation $\beta': G_1 \longrightarrow G_2$ by $\beta'(X) = \beta(S(C))$ for some choice of C such that $S(C) = X$. We claim β' is natural: In effect any $f: X \longrightarrow Y$ in Γ is the composite of the image in $\mathcal{X}$ of a finite sequence $(\theta_i)_{o\leq i\leq n}$ of arrows in $\mathcal{C}$ such that $X = S(\mathcal{S}(\theta_o))$, $Y = S(\mathcal{B}(\theta_n))$ and $\mathcal{B}(S(\theta_i)) = \mathcal{S}(S(\theta_{i+1}))$

thus for any such $f\colon X \to Y$ in Γ we have the commutative squares of the diagram

$$\begin{array}{ccccccccc}
G_1(Y) & \xrightarrow{G_1(S(\theta_n))} & G_1\mathcal{B}(S(\theta_{n-1}))) & \to \cdots \to & G_1\mathcal{B}(S(\theta_1)) & \xrightarrow{G_1S(\theta_0)} & G_1(X) \\
\big\downarrow{\scriptstyle \beta'(Y)} & & \big\downarrow{\scriptstyle \beta'\mathcal{B}S(\theta_{n-1}))} & & \big\downarrow{\scriptstyle \beta'\mathcal{B}S(\theta_1))} & & \big\downarrow{\scriptstyle \beta(X)} \\
G_2(Y) & \xrightarrow[G_2(S(\theta_n))]{} & G_2\mathcal{B}(S(\theta_{n-1}))) & \to \cdots \to & G_2\mathcal{B}(S(\theta_1)) & \xrightarrow[G_2S(\theta_0)]{} & G_2(X)
\end{array}$$

in which the composant squares are all commutative. Hence β' is a natural transformation and clearly $\beta' S = \beta$.

The lemma (5.10) is thus a consequence of the well known (and easily proved on the basis of Beck's theorem) fact that <u>given a commutative triangle</u>

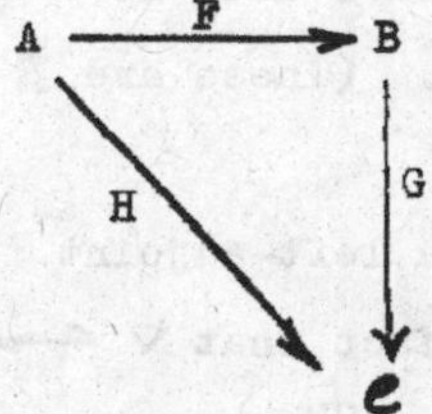

<u>of categories and functors in which G is a right-adjoint section</u>, H <u>is tripleable if and only if</u> F <u>is tripleable</u>.

Lemma (5.10) allows us to give an "absolute" criterion for tripleability over a functor category:

(5.11) <u>Proposition</u>. A category $\mathfrak{X}$ is tripleable over a functor category if and only if there exists a small subcategory $\mathcal{C}$ in $\mathfrak{X}$

such that the following conditions are satisfied:

1° $in : \mathcal{C} \hookrightarrow \mathfrak{X}$ is a co-frame;

2° $\mathfrak{X}$ has separators of pairs

3° for any $in_{\mathcal{C}}$-contractile equivalence pair $R \rightrightarrows X$ in $\mathfrak{X}$.

(a) $R \rightrightarrows X$ is effective and

(b) $R \rightrightarrows X \longrightarrow Y$ is bi-exact $\Longleftrightarrow$ for all $C \in \mathcal{O}b(\mathcal{C})$
$\mathfrak{X}(C,R) \rightrightarrows \mathcal{C}(C,X) \longrightarrow \mathfrak{X}(C,Y)$ is bi-exact.

(5.11.1) Example. (CAT) is tripleable relative to the subcategory ∇ of (CAT) whose objects consist of the categories Δ_0 and Δ_1 and whose non trivial arrows are the functors $d_0, d_1 : \Delta_0 \rightrightarrows \Delta_1$ defined by $0 \mapsto 1$ and $0 \mapsto 0$, respectively.

Δ_0 is, of course, the category whose only object is 0 and whose only arrow is the pair (0,0), while Δ_1 is the category whose objects consist of the integers 0 and 1 and whose arrows consist of the pairs (0,0), (1,1) and (0,1). (These are $\mathbb{1}$ and $\mathbb{2}$ in Lawvere's notation.)

In effect, the existence of a left-adjoint for (CAT) $\longrightarrow$ CAT(∇^{op},ENS), (i.e. the fact that $\nabla \hookrightarrow$ (CAT) is a co-frame) is well known: A functor $F: \nabla^{op} \longrightarrow$ ENS is just a diagram of sets of the form $X_1 \rightrightarrows X_0$ and the value of the left-adjoint at F is simply the category of paths of this diagram scheme (cf. GABRIEL-ZISMAN (1967)). (CAT) has separators so there only remains the verification of the existence and exactness conditions: Let $\mathcal{R} \overset{p_1}{\underset{p_2}{\rightrightarrows}} \mathcal{A}$ be a ∇-contractile equivalence pair. The statement that (p_1,p_2) is ∇-contractile says that there exist mappings s_1, s_0

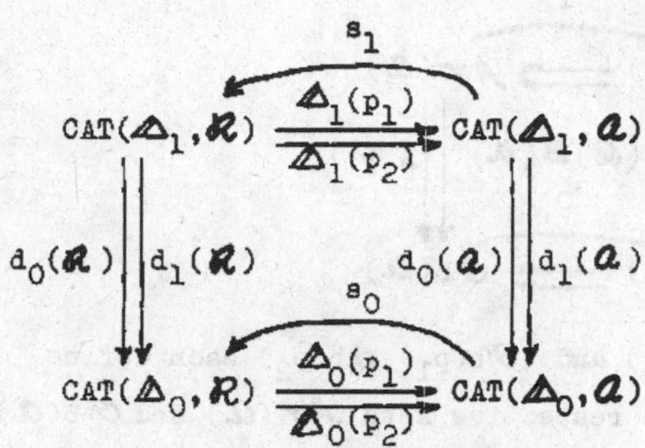

such that they are contractions for the equivalence pairs $(\Delta_1(p_1), \Delta_1(p_2))$ and $(\Delta_0(p_1),\Delta_0(p_2))$, respectively, and have the property that

$$d_0(R)s_1 = s_0 d_0(\mathcal{A}) \quad \text{and} \quad d_1(R)s_1 = s_0 d_1(\mathcal{A}) .$$

As is well known, for any category $\mathcal{X}$, $\mathrm{CAT}(\Delta_0,\mathcal{X})$ may be identified with $\mathcal{O}b(\mathcal{X})$ and $\mathrm{CAT}(\Delta_1,\mathcal{X})$ with $\mathcal{A}r(\mathcal{X})$, while the $d_0(\mathcal{X})$ and $d_1(\mathcal{X})$ may be identified with the co-domain and domain mappings for the category $\mathcal{X}$.

Similarly for any functor $F\colon \mathcal{X} \to \mathcal{Y}$, $\Delta_0(F)$ and $\Delta_1(F)$ are identifiable with the respective, object and arrow mappings of the functor F. Thus the mappings s_0 and s_1 merely define a domain and co-domain preserving mappings between the arrow sets of the categories $\mathcal{A}$ and $\mathcal{R}$ $\langle\!\langle S(A \xrightarrow{f} B) \mapsto S(A) \xrightarrow{S(f)} S(B)\rangle\!\rangle$ which are, of course, not necessarily associated with those of a functor.

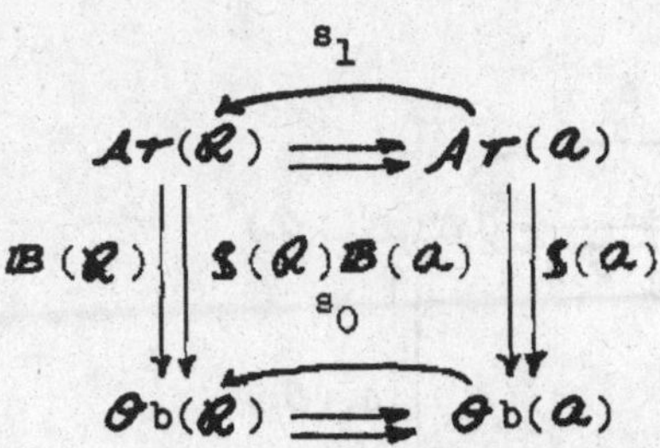

The pairs $(\mathcal{A}r(p_1), \mathcal{A}r(p_2))$ and $(\mathcal{O}b(p_1), \mathcal{O}b(p_2)$ each define equivalence relations on the respective sets $\mathcal{A}r(\mathcal{A})$ and $\mathcal{O}b(\mathcal{A})$. Let Q_1 and Q_0 be the respective quotients of these sets with ν_1 and ν_0 the canonical surjections, q_0 and q_1 the mappings deduced from $\mathbb{B}(\mathcal{A})$ and $\mathbb{S}(\mathcal{A})$ by passage to quotients, and t_1 and t_0 the distinguished sections of ν_1 and ν_0 which are obtained from s_1 and s_0. We thus obtain the sequentially commutative diagram

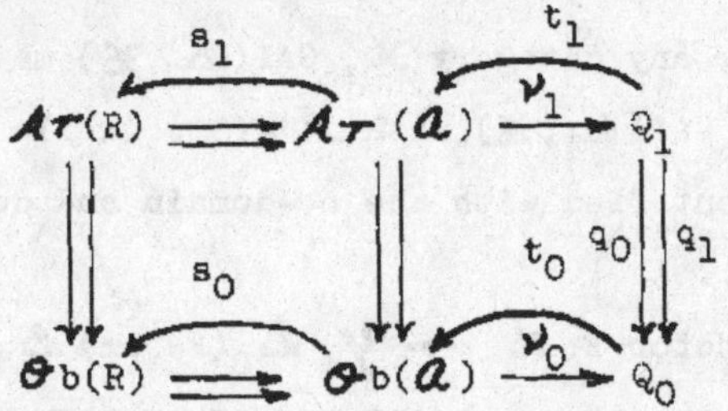

of sets and mappings whose lines are bi-exact contractile sequences. It is trivial to verify that given any such diagram, the induced diagram

$$\mathcal{A}r(\mathcal{R}) \times_{\mathbb{B},\mathbb{S}} \mathcal{A}r(\mathcal{R}) \rightrightarrows \mathcal{A}r(\mathcal{A}) \times_{\mathbb{B},\mathbb{S}} \mathcal{A}r(\mathcal{A}) \longrightarrow Q_1 \times_{q_0,q_1} Q_1$$

obtained by taking fiber-products over the base sequence, is bi-exact and contractile.

It is thus the case that the multiplication $\mu(\mathcal{A})$: $\mathcal{A}r(\mathcal{A}) \underset{\mathcal{B},\mathcal{S}}{\times} \mathcal{A}r(\mathcal{A}) \to \mathcal{A}r(\mathcal{A})$, by passage to quotients, defines a mapping $\mu: Q_1 \underset{q_0,q_1}{\times} Q_1 \to Q_1$.

Define now a category $\mathcal{Q}$ which has as arrows the set Q_1, and as objects the set Q_0 with multiplication the function μ. It is easy to verify that this is indeed a category and that with ν_1 as arrow mapping and ν_0 as object mapping one has a functor $\nu: \mathcal{A} \to \mathcal{Q}$ which is in fact a co-kernel for the pair (p_1,p_2) with (p_1,p_2) as its kernel pair and that in fact

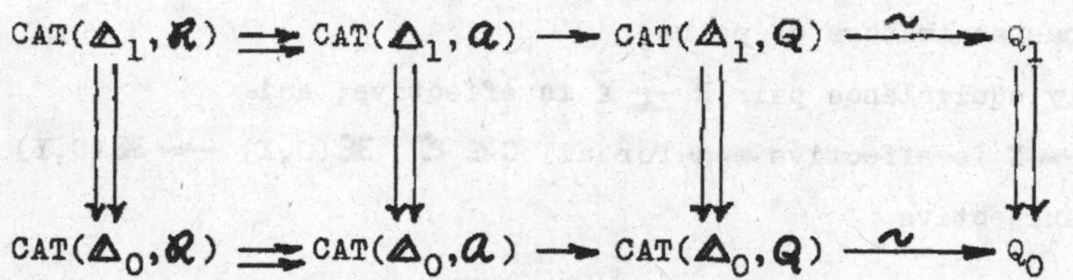

$$\begin{array}{ccccccc} CAT(\Delta_1,\mathcal{R}) & \rightrightarrows & CAT(\Delta_1,\mathcal{A}) & \to & CAT(\Delta_1,\mathcal{Q}) & \xrightarrow{\sim} & Q_1 \\ \downdownarrows & & \downdownarrows & & \downdownarrows & & \downdownarrows \\ CAT(\Delta_0,\mathcal{R}) & \rightrightarrows & CAT(\Delta_0,\mathcal{A}) & \to & CAT(\Delta_0,\mathcal{Q}) & \xrightarrow{\sim} & Q_0 \end{array}.$$

Moreover, any functor $\mathcal{A} \to \mathcal{Q}$ which equalizes p_1 and p_2 and projects on $Q_1 \rightrightarrows Q_0$ is itself a co-kernel for (p_0,p_1).

We have thus verified that (CAT) is tripleable relative to ∇(i.e. is ∇- tripleable).

REMARK. It is well known that (CAT) is a GABRIEL-retract of a functor category (and hence tripleable over same). Take $\hat{\mathcal{C}}$, where $\mathcal{C}$ is that full subcategory of (CAT) defined by the objects Δ_0,

Δ_1, Δ_2, and which looks like

together with the identity functors.

(5.12) Corollary. A category $\mathfrak{X}$ is tripleable over a functor category $\hat{\mathcal{C}}$ with $\mathcal{C}$ discrete if and only if there exists a small collection $\mathcal{C}$ of objects of $\mathfrak{X}$ such that the following conditions are satisfied.

1° The co-product of any small family $(X_i)_{i \in I}$ of objects $X_i \in \mathcal{C}$ exists in $\mathfrak{X}$;

2° $\mathfrak{X}$ has separators of pairs;

3° Every equivalence pair $R \rightrightarrows X$ is effective; and

4° $X \to Y$ is effective $\Longleftrightarrow$ for all $C \in \mathcal{C}$, $\mathfrak{X}(C,X) \to \mathfrak{X}(C,Y)$ is surjective.

(An interesting example of (5.12) will be discussed in (6.2).)

In particular (5.12) gives the following interesting

(5.13) Corollary. A category $\mathfrak{X}$ is tripleable over (ENS) if and only if there exists an object X_0 in $\mathfrak{X}$ such that

1° Any co-power of X_0 exists in $\mathfrak{X}$;

2° $\mathfrak{X}$ has separators of pairs;

3° every equivalence pair is effective; and

4° $f\colon X \to Y$ is effective $\Longleftrightarrow \mathfrak{X}(X_0,X) \to \mathfrak{X}(X_0,Y)$ is surjective.

In the usual cases X_0 is, of course, the free object on one generator, $\mathbb{Z}$ in the case of groups for instance. Moreover, X_0 is also a projective generator in the traditional sense of the term (but not in Gabriels' usage, of course). That is to say, in addition to carrying effective epimorphisms into surjections one also has that <u>for all $T \in \mathcal{O}b(\mathfrak{X})$ such that</u> T <u>is not isomorphic to co-terminal object $\emptyset$</u> (which must exist as a co-limit over the void category) <u>one has</u> $\mathfrak{X}(X_0,T) \neq \emptyset$.

In effect consider any $T \xrightarrow{\sim} \emptyset$ and suppose that $\mathfrak{X}(X_0,T) = \emptyset$. As one always has the unique arrow $\alpha: \phi \longrightarrow T$ consider $X_0(\alpha)$: $\mathfrak{X}(X_0,\phi) \longrightarrow \mathfrak{X}(X_0,T)$. As $\mathfrak{X}(X_0,T)$ is void by our supposition, then $\mathfrak{X}(X_0,\phi)$ must also be void, but then $X_0(\alpha)$ is a bijection, which implies that α must have been an isomorphism since $\mathfrak{X}$ was tripleable with respect to h'_{X_0}. This is a contradiction, hence we have the result. Of course, $\mathfrak{X}(X_0,\phi)$ may or, may not be empty. In the case of groups for example, $\phi \xrightarrow{\sim} 1$ and $X_0 \xrightarrow{\sim} \mathbb{Z}$ and $G_{\mathcal{R}}(\mathbb{Z},1) \neq \emptyset$, but in (ENS), $\phi \xrightarrow{\sim} \emptyset$ and $X_0 \xrightarrow{\sim} \{\emptyset\}$ and $\text{ENS}(\{\emptyset\},\emptyset) = \emptyset$.

This corollary parallels Linton's generalization of Lawvere's theories (Lawvere 1963) and shows that the only essential difference for infinitary operations is the absence of an "<u>abstractly finite</u>" projective generator, which is, of course, exactly what one would expect! Our version of this theorem, of course, could have been obtained directly from Linton's theorem (4.4 here). As the work here shows, it is unnecessary to verify the existence of arbitrary co-kernels. Their only purpose (as far as hypotheses go) seems to be to guarantee the existence of separators (i.e. an <u>inverse</u>

limit, c.f. (1.3)). Placing this here simply fits tripleability over (ENS) into the more general phenomenon of tripleability over functor categories by means of what Isbell has called a left-regular representation (Isbell 1961).

(5.14) Another easily made observation is that if $\mathfrak{X}$ is tripleable over (ENS) then <u>every effective epimorphism is universal</u>, where by <u>universal effective</u> we mean that the fiber product of the epimorphism with any possible arrow in the category exists and that the resulting arrow (pr_1) obtained by such a "change of base" is itself an effective epimorphism.

This latter is an immediate consequence of the fact that any functor U: $\mathcal{A} \to \mathcal{B}$ which generates fiber products and satisfies the condition (R) of (3.3) automatically satisfies the stronger condition:

(R*) for all $f \in \mathcal{A}r(\mathcal{A})$, U(f) <u>a squareable retraction</u> $\Rightarrow$ f <u>is a universal effective epimorphism</u> (by <u>squareable</u> we mean that any possible fiber product over $\mathbb{B}(U(f))$ exists in $\mathcal{B}$).

In effect, consider g: X $\to \mathbb{B}(f)$, the fiber product $U(X) \underset{U(g),U(f)}{\times} U(\mathbb{S}(f))$ exists, hence the fiber product $X \underset{g,f}{\times} \mathbb{S}(f)$ exists in $\mathcal{A}$ and the commutative square

$$\begin{array}{ccc} U(X \underset{g,f}{\times} \mathbb{S}(f)) & \xrightarrow{U(pr_2)} & U(\mathbb{S}(f)) \\ \downarrow{\scriptstyle U(pr_1)} & & \downarrow{\scriptstyle U(f)} \\ U(X) & \xrightarrow[U(g)]{} & U(\mathbb{B}(f)) \end{array}$$

is cartesian, if $U(f)$ is a retraction, then so is $U(pr_1)$ since the square is cartesian. Thus by (R), pr_1 must have been effective, i.e., f was a universal effective epimorphism.

Thus, if $\mathfrak{X}$ is tripleable over any "set like" category $\mathcal{B}$ (i.e., $\mathcal{B}$ satisfies the hypothesis of (4.1)), then every equivalence pair is universal and effective (i.e., its co-kernel is a universal effective epimorphism) and every effective epimorphism (and, thus, in particular, every co-kernel arrow) is universal effective. This latter has the effect of eliminating many interesting categories from consideration as tripleable over any "set like" category. For example, TopSp and CAT are not tripleable over (ENS) for any functor whatever, as both contain effective epimorphisms which are not universal. TopSp was, of course, already eliminated by (5.12) since it contains equivalence pairs which are not effective.

It is interesting to translate (5.13) into the statement that its <u>opposite category</u> be tripleable over (ENS). This gives the

(5.15) <u>Corollary</u>. A category $\mathfrak{X}$ is such that $\mathfrak{X}^{op}$ is tripleable over (ENS) if and only if there exists an object X_0^* in $\mathfrak{X}$ such that

1° Any power of X_0^* exists in $\mathfrak{X}$;

2° $\mathfrak{X}$ has co-separators of pairs;

3° every co-equivalence pair is effective, and

4° $f: X \longrightarrow Y$ is an effective monomorphism $\Longleftrightarrow \mathfrak{X}(Y,X_0^*) \longrightarrow \mathfrak{X}(X,X_0^*)$ is surjective.

There are numerous examples of this phenomenon which occur in the literature [see for example SEMADINI (1963)]. We restrict ourselves here to a few examples and will defer the detailed study which this has merited to a separate article in which we will discuss its proper place in a hierarchical classification of "axioms on generation" starting with faithfulness and passing through Isbell's notion of adequacy (Isbell 1961) to Gabriel's retracts.

(5.15.1) <u>Examples</u>. If ENS is the <u>category of sets</u>, then ENS^{op} is tripleable over ENS with the two element set $\underline{2} = \mathcal{O}b(\Delta_1)$ as the object X_0^*. The resulting functor into ENS is nothing more than the contravariant power set functor--which is its own left-adjoint. The resulting category of algebras over this triple is, of course, nothing more than the category of complete atomic boolean algebras and the resulting strict duality is the familiar one.

(5.15.2) If Ab is the category of abelian groups, then Ab^{op} is tripleable over ENS with the toroidal group $\mathbb{Q}/\mathbb{Z}$ as an object X_0^*.

(5.15.3) If COMP is the category of <u>compact Hausdorff spaces</u>, then $COMP^{op}$ is tripleable over the category of sets with the closed unit interval $[0,1]$ as an object X_0^*.

(5.16) Quite in general one can use (5.13) for example to give simple "characterization theorems" of the usual categories of algebraic structures: Since a co-T-algebra in any category may be identified with a lifting of some hom-functor, in order that a given category $\mathfrak{X}$ be equivalent to a category of T-algebras, it is necessary and sufficient that it be tripleable over ENS by means of a generator which is a co-T-algebra in such a fashion that the defined triples co-incide. For example, in order that a category $\mathfrak{X}$ be equivalent to the category of groups, it is necessary and sufficient that $\mathfrak{X}$ satisfy the conditions of (5.13) for an object P which is a co-group such that as groups $\mathfrak{X}(P,S\cdot P) \xrightarrow{\sim} S\cdot \mathbf{Z}$ for any set S. Then $\mathfrak{X}(P_1P) \xrightarrow{\sim} \mathbf{Z}$ and as groups, $\mathfrak{X}(P,S\cdot P) \xrightarrow{\sim} S\cdot\mathfrak{X}(P_1P)$, etc. etc. This technique and its variants are used repeatedly.

(5.17) It follows from a theorem of Giraud (c.f. Verdier 1963) that in order that a U-category $\mathfrak{X}$ be equivalent to the category of sheaves over some $\mathcal{U}$-small-site (i.e., small category supplied with a Grothendieck topology), it is necessary and sufficient that $\mathfrak{X}$ have (a) (U-)coproducts which are disjoint and universal (for all terminology here see the reference of Verdier cited above), (b) finite limits, (c) every equivalence pair effective and universal, and (d) a family of $(X_i)_{i\in I}$ objects (indexed by $I\in\mathcal{U}$) for which any monomorphism

$$u: G \hookrightarrow H \text{ in } \mathfrak{X} \text{ is an isomorphism if and only if for all } i\in I$$
$$\mathfrak{X}(X_i,u): \mathfrak{X}(X_i,G) \longrightarrow \mathfrak{X}(X_i,H) \text{ is a bijection.}$$

It is not difficult to prove the above conditions are almost equivalent to the conditions of (5.12) together with the addition that co-products be disjoint and universal. For example, if $\mathfrak{X}$ has square fiber products and every equivalence pair is effective <u>and</u> universal, then any arrow $f\colon A \longrightarrow B$ factors through an effective epimorphism and a monomorphism

$$\mathbb{R}(f) \rightrightarrows A \xrightarrow{f} B, \quad A \xrightarrow{\nu} Q \xrightarrow{m} B$$

so that f is effective if and only if m is an isomorphism. Consequently, one may replace the condition (d) by «$f\colon A \longrightarrow B$ is effective provided for all $i \in I$

$$\mathfrak{X}(X_i,A) \longrightarrow \mathfrak{X}(X_i,B) \text{ is surjective} ».$$

The remarks of (5.14) then allow one to go in the other direction so that a $\mathcal{V}$-topos (i.e., $\mathcal{V}$-category which satisfies (a)-(d)) may be viewed as a category which is a retract of its category of algebras over a functor category $\mathcal{C}^{\wedge}$ with $\mathcal{C}$ discrete in which co-products are disjoint and universal. Giraud's theorem may be viewed as a modified "tripleability theorem" since the disjointness and universality of co-products are exactly what is required in order that the canonical functor of $\mathfrak{X}$ into $\mathcal{C}^{\sim}$, the category of sheaves on the full subcategory $\mathcal{C}$ defined by $(X_i)_{i \in I}$ supplied with the restriction of the canonical topology on $\mathfrak{X}$ to $\mathcal{C}$, commute with co-products, i.e., that the defining triples co-incide. This will be discussed fully in a separate article.

The proof given in the next section may, in fact, be viewed as a special case of Giraud's theorem where the topology in question is chaotic. We discuss it separately since it is elementary and requires no auxiliary definitions.

6. Equivalence with functor categories

We now use the preceding theorems to give alternatives to Linton's (1967) formulation of a theorem due to M. Bunge (1966) (see also Gabriel (1966)) characterizing those categories $\mathcal{X}$ which are equivalent to categories of the form $\hat{\mathcal{C}}$ for some small category $\mathcal{C}$. The idea of proof is due to Linton and Beck and is based on the following observations.

(6.1) Let $\mathcal{C}$ be a small category and $DIS(\mathcal{C})$ the discrete category associated with $\mathcal{C}$, with $in_D\colon DIS(\mathcal{C}) \to \mathcal{C}$ the inclusion functor. in_D induces a functor $\hat{in_D}\colon \hat{\mathcal{C}} \to DIS(\mathcal{C})\hat{}$ by restriction. Moreover $\hat{in_D}$ admits a left-adjoint which may be defined by $\ll G \mapsto \coprod_{X\in Ob(\ )} G(X)\circ h_X \gg$ since, as always, one has the chain of natural isomorphisms

$$DIS(\mathcal{C})\hat{}(G,\underline{F}) = \prod_{X\in Ob(\mathcal{C})} ENS(G(X),\underline{F}(X)) \xrightarrow{\sim} \prod_{X\in Ob(\mathcal{C})} \underline{F}(X)^{G(X)} \xrightarrow{\sim}$$

$$\prod_{X\in Ob(\mathcal{C})} \hat{\mathcal{C}}(h_X,F)^{G(X)} \longrightarrow \prod_{X\in Ob(\mathcal{C})} \hat{\mathcal{C}}(G(X)\circ h_X,F) \xrightarrow{\sim} \hat{\mathcal{C}}(\coprod_{X\in Ob(\ )} G(X)\circ h_X,F)$$

where $\underline{F} = F\, in_D$ for any $F\colon \mathcal{C}^{op} \to (ENS)$.

In addition $DIS(\mathcal{C})\hat{}$ has separators of pairs and every equivalence pair is effective and contractile. As every equivalence pair in $\hat{\mathcal{C}}$ is effective and $R \rightrightarrows S \to T$ is bi-exact if and only if $\underline{R} \rightrightarrows \underline{S} \to \underline{T}$ is bi-exact. The fact that $\hat{\mathcal{C}}$ has separators allows us to conclude by (4.2) that $\underline{\hat{\mathcal{C}} \xrightarrow{\hat{in}_D} DIS(\mathcal{C})^{\wedge} \text{ is tripleable}}$. As $\hat{\mathcal{C}}$ contains $\mathcal{C}$ as a full subcategory, in order that any category $\mathfrak{X}$ be equivalent to $\hat{\mathcal{C}}$ it will be necessary and sufficient that $\mathfrak{X}$ contain $\mathcal{C}$ as a full subcategory and that $in_D\, \hat{in}_{\mathcal{C}}\, h_{\mathfrak{X}}$: $\mathfrak{X} \to DIS(\mathcal{C})^{\wedge}$ be tripleable for the triple defined by $\hat{in}_{\mathcal{C}}$ and its left-adjoint. Consequently, we have the

(6.2) Proposition. In order that a category $\mathfrak{X}$ be equivalent to a category of the form $\hat{\mathcal{C}}$ for some small category $\mathcal{C}$ it is necessary and sufficient that $\mathfrak{X}$ contain a small set $\underline{\mathcal{C}}$ of objects such that

1° arbitrary small co-products of families of objects of $\underline{\mathcal{C}}$ exist;

2° $\mathfrak{X}$ has separators of pairs;

3° $\hat{in}_D \hat{in}_{\mathcal{C}}\, h_{\mathfrak{X}}$ verifies any one of the equivalent conditions of (4.2) (e.g. Every equivalence pair in $\mathfrak{X}$ is effective and for complexes $R \rightrightarrows S \to T$ in $\mathfrak{X}$, $R \rightrightarrows S \to T$ is bi-exact $\iff$ for all $C \in \mathcal{O}b(\mathcal{C})$, $\mathfrak{X}(C,R) \rightrightarrows \mathfrak{X}(C,S) \to \mathfrak{X}(C,T)$ is bi-exact, where $\mathcal{C}$ here denotes the full subcategory of $\mathfrak{X}$ defined by the set $\underline{\mathcal{C}}$).

4° for any small co-product $\coprod_{i \in I} C_i$ in $\mathfrak{X}$ with $C_i \in \underline{\mathcal{C}}$ for all $i \in I$,

$$\mathfrak{X}(C, \coprod_{i \in I} C_i) \xrightarrow{\sim} \coprod_{i \in I} \mathfrak{X}(C,C_i) \text{ for all } C \in \underline{\mathcal{C}}.$$

In effect 1°-3° guarantee that $\hat{in}_D \hat{in}_{\mathcal{C}} h_{*}$ is tripleable and 4° guarantees that the triple is equivalent to that of $\hat{in}_D$. Q.E.D.

7. Remarks on the existence of co-kernels

(7.0) Let $A \underset{y}{\overset{x}{\rightrightarrows}} B$ be a double arrow in some category $\mathcal{A}$ which has square fiber products. Suppose that the pair (x,y) admits a co-kernel $\nu: B \to Q$. Then $\mathcal{R}(\nu) \rightrightarrows B$ is an effective equivalence pair for which there exists a (unique) arrow $\langle x,y\rangle: A \to \mathcal{R}(\nu)$ such that $pr_1\langle x,y\rangle = x$ and $pr_2\langle x,y\rangle = y$. In addition $\mathcal{R}(\nu) \rightrightarrows B$ defines a representation of the functor defined by $\langle\langle T \mapsto \mathcal{R}(\langle \mathcal{A}(T,f)\rangle_{f \in \mathcal{A}r(x,y)})\rangle\rangle$ where

$$\mathcal{R}(\langle \mathcal{A}(T,f)\rangle_{f \in \mathcal{A}r(x,y)}) \rightrightarrows \mathcal{A} \to \prod_{f \in \mathcal{A}r(x,y)} \mathcal{A}(T,\mathcal{B}(f))$$

is exact and $\mathcal{A}r(x,y)$ is the (not necessarily $\mathcal{U}$- small) set of all $f \in \mathcal{A}r(\mathcal{A})$ such that $\mathcal{S}(f) = B$ and $fx = fy$.

Conversely, if this last functor is representable and its representative admits a co-kernel, this same representative is an effective equivalence pair and its co-kernel is also a co-kernel for the pair (x,y). As $\mathcal{R}(\langle \mathcal{A}(T,f)\rangle_{f \in \mathcal{A}r(x,y)})$ is bijectively equivalent to the intersection $\bigcap_{f \in \mathcal{A}r(x,y)} \mathcal{R}(\mathcal{A}(T,f)) \xrightarrow{\sim} \bigcap_{f \in \mathcal{A}r(x,y)} \mathcal{A}(T,\mathcal{R}(f))$ we say in this case that the <u>intersection of the family of equivalence pairs</u> $(\mathcal{R}(f))_{f \in \mathcal{A}r(x,y)}$ <u>exists and is effective</u>.

(7.1) Consider now for such a pair (x,y) the following categories:

EqCo(B): the category of equivalence pairs of the form $R \rightrightarrows B$ (with arrows those in $\mathcal{A}$ which commute with the projections).

$\mathcal{R}(\mathcal{A}^{op}/B)$: the category of kernel pairs i.e., the full subcategory of EqCo(B) consisting of those equivalence pairs each of which have the form $\mathcal{R}(f) \rightrightarrows B$ for some $f: B \to T$.

$\mathcal{E}(B)$: the full subcategories of $\mathcal{R}(\mathcal{A}^{op}/B)$ consisting of the effective equivalence pairs above B, and finally the full subcategory,

$\mathcal{R}_{(x,y)}(\mathcal{A}^{op}/B)$ of equivalence pairs for which there exists an arrow $s: A \to R \underset{pr_2}{\overset{pr_1}{\rightrightarrows}} B$ such that $pr_1 s = x$ and $pr_2 s = y$.

One then has the following inclusions

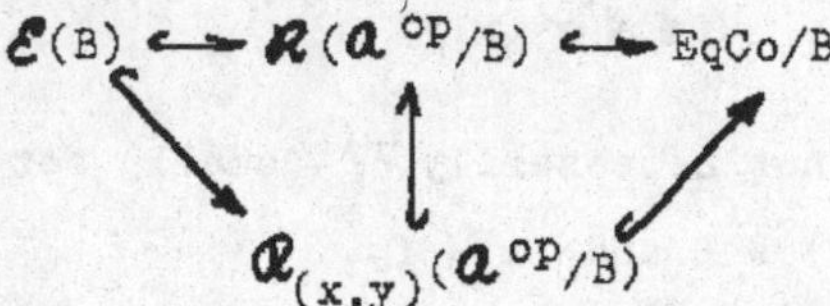

and the elementary observation that (x,y) <u>admits a co-kernel</u> iff <u>the category</u> $\mathcal{R}_{(x,y)}(\mathcal{A}^{op}/B)$ <u>has an initial object which is effective</u>.

(7.2) <u>Definition</u>. Let $\| EqCo(B) \|$ be the category of isomorphism classes of equivalence pairs in $\mathcal{A}$ above the object B in $\mathcal{A}$. We say that $\mathcal{A}$ is <u>equivalence-small</u> if for any $B \in \mathcal{O}b(\mathcal{A})$ the category $\| EqCo(B) \|$ is equivalent to a ($\mathcal{N}$-) small category, and <u>equivalence algebraic</u> provided $\mathcal{A}$ is equivalence small and the intersection of any small family of members of EqCo(B) exists, i.e. the functor $\ll T \mapsto \bigcap_{\gamma \in I} \langle pr_1(T), pr_2(T)\rangle \langle \mathcal{A}(T, \mathcal{R}_\gamma) \rightarrowtail \mathcal{A}(T,B) \times \mathcal{A}(T,B) \gg$ is representable for any $I \in \mathcal{M}$.

(7.3) Proposition. Let $U: \mathcal{A} \to \mathcal{B}$ be a functor which satisfies the condition of Lemma (3.4). Then if $\mathcal{B}$ is equivalence small, so is $\mathcal{A}$.

In effect the square (D) which occurs in the proof of (3.4) is cartesian, so that equivalence pairs which are isomorphic under U must themselves be isomorphic.

(7.4) Corollary. If $U: \mathcal{A} \to \mathcal{B}$ is tripleable, then if $\mathcal{B}$ is equivalence-algebraic, so is $\mathcal{A}$.

This is trivial since U generates projective limits.

(7.5) Definition. An equivalence pair $R \underset{pr_2}{\overset{pr_1}{\rightrightarrows}} B$ in $\mathcal{A}$ is said to be strict provided that for any double arrow $T \underset{y}{\overset{x}{\rightrightarrows}} B$ the condition

$\ll$for all $f: B \to S$, $f\, pr_1 = f\, pr_2 \Rightarrow fx = fy \gg$

which is necessary for (x,y) to factorize through (p_1,p_2), is also sufficient for such a factorization, i.e. if $\mathcal{A}r(p_1,p_2)$ is the set of all $f: B \to B(f)$ such that $fp_1 = fp_2$ the sequence

$$\mathcal{A}(T,R) \hookrightarrow \mathcal{A}(T,B) \times \mathcal{A}(T,B) \rightrightarrows \prod_{f \in A\tau(p_1,p_2)} \mathcal{A}(T,\mathcal{B}(f))$$

defined in the obvious manner is exact.

Most of the equivalence pairs which we have met here, e.g. kernel pairs, separators, intersections of families of kernel pairs, etc. are strict.

(7.6) <u>Proposition</u>. If $\mathcal{A}$ has square fiber products and every strict equivalence pair is semi-effective (and hence effective), then if $\mathcal{A}$ is equivalence-algebraic, any double arrow $A \overset{x}{\underset{y}{\rightrightarrows}} B$ for which there exists an $f: B \to S$ such that $fx = fy$ has a co-kernel.

In effect we take the co-kernel of the intersection of a system of representatives of the kernel pairs of all those arrows which equalize (x,y).

(7.7) <u>Proposition</u>. If $U: \mathcal{A} \to \mathcal{B}$ is a tripleable functor into a category with square fiber products and a terminal object in which every equivalence pair is contractile and semi-effective, and which is moreover equivalence-algebraic, then

1° $\mathcal{A}$ has co-kernels, and

2° $\mathcal{B}$ has co-products $\Rightarrow \mathcal{A}$ has co-products.

We simply apply (2.6) in its lifted version to the pairs (x,y). <u>Notice that this does not require that</u>

$\ll A \rightrightarrows B \to C$ co-exact $\Rightarrow U(A) \rightrightarrows U(B) \to U(C)$ co-exact $\gg$

<u>for arbitrary pairs, however</u>.

For the existence of co-products we note that for any family $(X_i)_{i\in I}$ in $\mathcal{A}$, the canonical sequence

$$FU(\mathcal{R}(\beta_i)) \rightrightarrows FU(X_i) \longrightarrow X_i$$

is co-exact for each $i \in I$. Hence the sequence

$$\prod_{i\in I} \mathcal{A}(X_i,T) \hookrightarrow \prod_{i\in I} \mathcal{A}(FU(X_i),T) \rightrightarrows \prod_{i\in I} \mathcal{A}(FU(\mathcal{R}(\beta_i),T)$$

is exact so that the sequence

$$\prod_{i\in I} \mathcal{A}(X_i,T) \hookrightarrow \mathcal{A}(F(\coprod_{i\in I} U(X_i),T) \rightrightarrows \mathcal{A}(F(\coprod_{i\in I} U(\mathcal{R}(\beta_i)),T)$$

is exact so that $F(\coprod_{i\in I} U(\mathcal{R}(\beta_i))) \rightrightarrows F(\coprod_{i\in I} U(X_i) \longrightarrow \coprod_{i\in I} X_i$ is co-exact.

(7.8) Corollary. If $U: \mathcal{A} \longrightarrow \mathcal{B}$ is a functor which verifies the conditions of (7.7) then $\mathcal{B}$ has co-limits $\Rightarrow \mathcal{A}$ has co-limits.

(7.9) Corollary. If $U_1: \mathcal{A}_1 \longrightarrow \mathcal{B}$ and $U_2: \mathcal{A}_2 \longrightarrow \mathcal{B}$ are tripleable functors into a category $\mathcal{B}$ with square fiber products and which is equivalence-algebraic, and in which every equivalence pair is contractile and semi-effective, then any functor $S: \mathcal{A}_1 \longrightarrow \mathcal{A}_2$ such that $U_2S \xrightarrow{\sim} U_1$ admits a left-adjoint and is tripleable.

By (7.4) $\mathcal{A}_1$ is equivalence-algebraic and by (4.2) every equivalence couple is effective. Thus, to construct a left-adjoint it will suffice to take as its value at $X \in \mathcal{O}b\,\mathcal{A}_2$ the co-kernel of the intersection of a system of representatives of those equivalence pairs associated with those arrows $f: F_1U_2(X) \to Y$ which arise as values under the image of the transformation

$$\mathcal{A}_2(X,S(Y)) \hookrightarrow \mathcal{B}(U_2(X),U_2S(Y)) \xrightarrow{\sim} \mathcal{B}(U_2(X),U_1(Y)) \xrightarrow{\sim} \mathcal{A}_1(F_1U_2(X),Y)$$

where F_1 is the given left-adjoint for U_1. The verification is elementary. For the other part it is well known that if U_1 and U_2 are tripleable, then S if tripleable if and only if it admits a left adjoint.

(6.10) Corollary. The above corollary (7.9) is valid where B is (ENS) [Linton (1965)], and more generally whenever B is of the form $\hat{\mathcal{C}}$ for some small discrete category $\mathcal{C}$.

(If $\mathcal{C}$ is a small discrete category then $\hat{\mathcal{C}}$ is equivalence small since for any $B \in \mathcal{O}b(\hat{\mathcal{C}})$, $\mathrm{Card}\,\|\mathrm{EqCo}(B)\| < \mathrm{Card}(\prod_{c \in \mathcal{O}b(\mathcal{C})^{\beta}} (B(c) \times B(c)) \leq x \in \mathcal{V}\!\!\mathcal{L}$.)

References

1. BECK, J., untitled manuscript, Cornell, 1966.
2. BUNGE, M., Relative Functor Categories and Categories of Algebras. Journal of Algebra Vol. II, Jan. 1969, pp. 64-101.
3. BUNGE, M., Dissertation. Univ. of Penn. 1966.
4. GABRIEL, P., handwritten draft (1966) of § 2 of Chevaley and Gabriel, Catégories et foncteurs (to appear).
5. GABRIEL, P., and ZISMAN, M., Calculus of Fractions and Homotopy Theory. Springer, Berlin, 1967.
6. GROTHENDIECK, A., Techniques de construction et théorèmes d'existence en géométrie algébrique, III: Préshémas quotients. Seminaire Bourbaki no. 212 (Feb. 1961) [see also Sem. Bourbaki 190 (Dec. 1959)].
7. ISBELL, J., Subobjects, adequacy, completeness and categories of algebras. Rozprawy Matematyczne XXXVI, Warsaw 1964.
8. LAWVERE, F. W., Functorial Semantics of Algebraic Theories. Proc. N.A.S. Vol. 50, 1963, page 869.
9. LINTON, F.E.J., Some Aspects of Equational Categories. Proc. Conf. Categ. Alg. (La Jolla, 1965), Springer, Berlin, 1966, pp. 84-94.
10. LINTON, F.E.J., Applied Functional Semantics, II. E.T.H. Triples Seminar Notes, 1967. Springer Lecture Notes #80 (Berlin 1969).
11. SEMADENI, Z., Projectivity, Injectivity, and Duality. Rozprawy Matematyczne XXXV, Warsaw, 1963.
12. VERDIER, J. L., Cohomologie Etale des Schemas, Séminaire de géometrie algébrique (S.G.A.A.) 1963-64. Fascicule 1. I.H.E.S. (1964).

AUTONOMOUS CATEGORIES WITH MODELS

by Myles Tierney[(1)]

Received April 23, 1969

0. Introduction

In [1] Appelgate and I defined a category with models to be a functor

$$I: \underline{M} \longrightarrow \underline{A}$$

where $\underline{M}$ --the model category--was small and $\underline{A}$ was arbitrary. In examples, however, it often occurs that $\underline{A}$ is an autonomous category in the sense of Linton [8] (the definition will be recalled below). In this case, if $\underline{A}$ has small limits, the singular functor

$$s: \underline{A} \longrightarrow (\underline{M}^*, S)$$

of [1] can be lifted to a strong functor

$$s: \underline{A} \longrightarrow (\underline{M}^*, \underline{A})$$

which, it will be shown, has a strong coadjoint if $\underline{A}$ has small colimits. Thus one obtains a strong model-induced cotriple $\mathbb{G}$ on $\underline{A}$. Since $\mathbb{G}$ is strong, it follows that the category $\underline{A}_{\mathbb{G}}$ of $\mathbb{G}$-coalgebras has a natural $\underline{A}$-structure, and there is a strong adjoint pair

$$\underline{A}_{G} \underset{\overline{r}}{\overset{\overline{s}}{\rightleftarrows}} (\underline{M}^*, \underline{A})$$

---

(1) The author was partially supported by the N.S.F. under Grant GP 8618.

Finally, one can extend the theorem of [1] which gives necessary and sufficient conditions for this pair to be an equivalence of categories.

Several examples are considered in §2, one of which gives a proof of the theorem of Dold and Kan that the category of FD-modules is equivalent to the category of positive chain complexes. I would like to thank H. Appelgate for the example of torsion groups, which he originally suggested in connection with [1].

I should remark that everything done here can be generalized to the setting of Eilenberg and Kelley [5], where the underlying set functor U: $\underline{A} \longrightarrow S$ is not assumed to be faithful. (In fact, some of this has been carried out by M. Bunge in [2].) However, the results became more complicated then, and since U is faithful in most of the examples, I have chosen to work with the simpler definitions of Linton rather than in the more general setting of [5].

I shall assume throughout a basic knowledge of cotriples such as can be found, for example, in [6]. Also, some acquaintance with [1] would be desirable, though it is not strictly necessary.

Recall that an autonomous category $\underline{A}$ is a category satisfying the axioms A1 - A5 of Linton [8]. That is, we require first that there be functors

$$U: \underline{A} \longrightarrow S$$

and

$$(-,-): \underline{A}^* \times \underline{A} \longrightarrow \underline{A}$$

such that U is faithful and

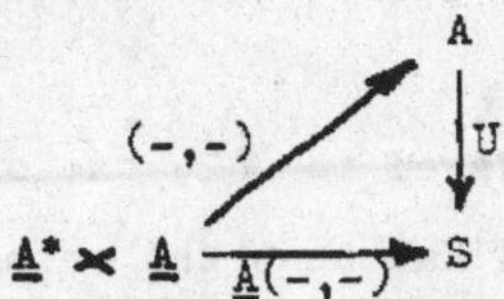

commutes--this data constitutes A1 - A3. For A4 there should exist, for each $A \in \underline{A}$, a strong coadjoint

$$-\otimes A\colon \underline{A} \longrightarrow \underline{A}$$

to the functor

$$(A,-)\colon \underline{A} \longrightarrow \underline{A} \;,$$

in the sense that there are natural $\underline{A}$-isomorphisms

$$(-\otimes A,-) \xrightarrow{\sim} (-,(A,-))$$

Finally, A5 implies that for all $A, B \in \underline{A}$ there are isomorphisms

$$A \otimes B \xrightarrow{\sim} B \otimes A$$

which are natural in A and B. We refer to Linton Proposition 2.4 for the fact that such isomorphisms insure that for each $A \in \underline{A}$ the contravariant functor

$$(-,A)\colon \underline{A} \longrightarrow \underline{A}$$

is strongly adjoint to itself on the right, and hence takes colimits into limits.

If $\underline{A}$ is an autonomous category, a category $\underline{B}$ is called an $\underline{A}$-category if there is given a functor

$$(-,-): \underline{B}^* \times \underline{B} \longrightarrow \underline{A}$$

such that

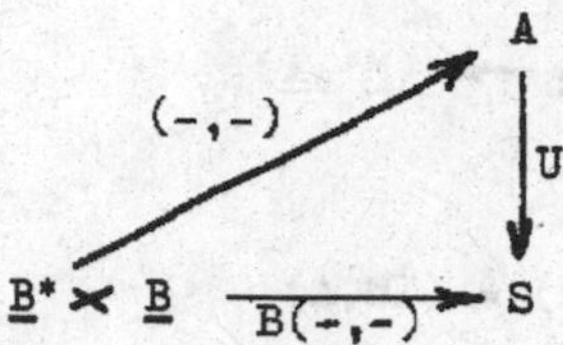

commutes, and also a strong composition

$$(B,C) \longrightarrow ((D,B),(D,C))$$

in $\underline{A}$ whose underlying is the ordinary composition of $\underline{B}$.

If $\underline{B}$ and $\underline{C}$ are $\underline{A}$-categories, a functor $F: \underline{B} \longrightarrow \underline{C}$ is said to be strong if the function

$$F: \underline{B}(B,B') \longrightarrow \underline{C}(FB,FB')$$

is the underlying of an $\underline{A}$-morphism

$$(B,B') \longrightarrow (FB,FB')$$

for each pair B, B' in $\underline{B}$.

1. Autonomous categories with models.

Throughout this section, $\underline{A}$ will denote an autonomous category with small limits and colimits, and we will assume U preserves the former.

Now let $\underline{M}$ be a small category and

$$I: \underline{M} \to \underline{A}$$

a functor -- we make no strongness assumptions on I or $\underline{M}$. Then I determines a functor

$$s: \underline{A} \to (\underline{M}^*, \underline{A})$$

given by

$$sA(M) = (IM, A)$$

for $A \in \underline{A}$ and $M \in \underline{M}$. The effect of s on morphisms is given by the functoriality of (-,-) in both variables.

The existence of small limits in $\underline{A}$ preserved by U makes $(\underline{M}^*, \underline{A})$ an $\underline{A}$-category by requiring that the following diagram be an equalizer in $\underline{A}$ for all $F_1, F_2 \in (\underline{M}^*, \underline{A})$:

$$(F_1, F_2) \to \prod_{M \in \underline{M}} (F_1 M, F_2 M) \overset{\Upsilon_1}{\underset{\Upsilon_2}{\rightrightarrows}} \prod_{\alpha: M'' \to M'} (F_1 M', F_2 M'')$$

where
$$P_\alpha \circ \Upsilon_1 = (F_1 \alpha, F_2 M'') \circ P_{M''}$$
$$P_\alpha \circ \Upsilon_2 = (F_1 M', F_2 \alpha) \circ P_{M'}$$

Now, with this $\underline{A}$-structure we construct a strong coadjoint r to s. That is, we will define a functor

$$r: (\underline{M}^*, \underline{A}) \to \underline{A} \quad ,$$

and produce, for each $A \in \underline{A}$ and $F \in (\underline{M}^*, \underline{A})$, an A-isomorphism

$$(rF,A) \longrightarrow (F,sA)$$

natural in A and F. We do this by requiring that for each $F \in (\underline{M}^*,\underline{A})$ the following be a coequalizer in $\underline{A}$:

$$\coprod_{\alpha:M'' \to M'} FM' \otimes IM'' \underset{\tau_2}{\overset{\tau_1}{\rightrightarrows}} \coprod_{M \in \underline{M}} FM \otimes IM \longrightarrow rF$$

where
$$\tau_1 \cdot i_\alpha = i_{M''} \circ F\alpha \otimes IM''$$
$$\tau_2 \cdot i_\alpha = i_{M'} \circ FM' \otimes I\alpha .$$

By a previous remark, when we apply $(-,A)$ to this coequalizer we obtain an equalizer

$$(rF,A) \longrightarrow (\coprod_{M \in \underline{M}} FM \otimes IM, A) \underset{(\tau_2,A)}{\overset{(\tau_1,A)}{\rightrightarrows}} (\coprod_{\alpha:M'' \to M'} FM' \otimes IM'', A)$$

where
$$(i_\alpha, A) \circ (\tau_1, A) = (F\alpha \otimes IM'', A) \circ (i_{M''}, A)$$
$$(i_\alpha, A) \circ (\tau_2, A) = (FM' \otimes I\alpha, A) \circ (i_{M'}, A).$$

Now $(-,A)$ takes coproducts to products, and making this identification we obtain an equalizer

$$(rF,A) \longrightarrow \prod_{M \in \underline{M}} (FM \otimes IM, A) \underset{\tau_2'}{\overset{\tau_1'}{\rightrightarrows}} \prod_{\alpha:M'' \to M'} (FM' \otimes IM'', A)$$

with τ_1' and τ_2' given by

$$P_\alpha \circ \tau_1' = (F\alpha \otimes IM'', A) \circ P_{M''}$$

$$P_\alpha \circ \tau_2' = (FM' \otimes I\alpha, A) \circ P_{M'} \quad .$$

Finally, using the strong hom-$\otimes$ adjointness we obtain an equalizer

$$(rF,A) \longrightarrow \prod_{M \in \underline{M}} (FM,(IM,A) \underset{\tau_2''}{\overset{\tau_1''}{\rightrightarrows}} \prod_{\alpha : M'' \to M'} (FM',(IM'',A))$$

where $\quad P_\alpha \circ \tau_1'' = (F\alpha,(IM'',A)) \circ P_{M''}$

$$P_\alpha \circ \tau_2'' = (FM',(I\alpha,A)) \circ P_{M'} \quad .$$

However, the equalizer of this diagram is, by definition, (F,sA), so that r is strongly coadjoint to s. We recall [8], Lemma 2.1, that this implies both r and s are strong. Thus, if

$$\eta : (\underline{M}^*,\underline{A}) \longrightarrow sr$$

and

$$\varepsilon : rs \longrightarrow \underline{A}$$

denote, respectively, the unit and counit of the adjointness, then

$$\mathbf{G} = (G,\varepsilon,\delta) = (rs,\varepsilon,r\eta s)$$

is a strong cotriple in $\underline{A}$, i.e., a cotriple whose functor part is strong. We call $\mathbf{G}$ the <u>strong model induced cotriple</u>.

For later use, we give a closer description of ε and η. First, if

$$\rho : \coprod_{M \in \underline{M}} FM \otimes IM \longrightarrow rF$$

denotes the natural projection in the coequalizer defining rF, let j_M be the composite

$$FM\otimes IM \xrightarrow{i_M} \coprod_{M\in \underline{M}} FM\otimes IM \xrightarrow{\rho} rF \quad .$$

Then a morphism f: rF $\longrightarrow$ A is given by a family

$$f_M = f\circ j_M\colon FM\otimes IM \longrightarrow A,$$

indexed by $M \in \underline{M}$, such that for each $\alpha\colon M'' \longrightarrow M'$ in $\underline{M}$,

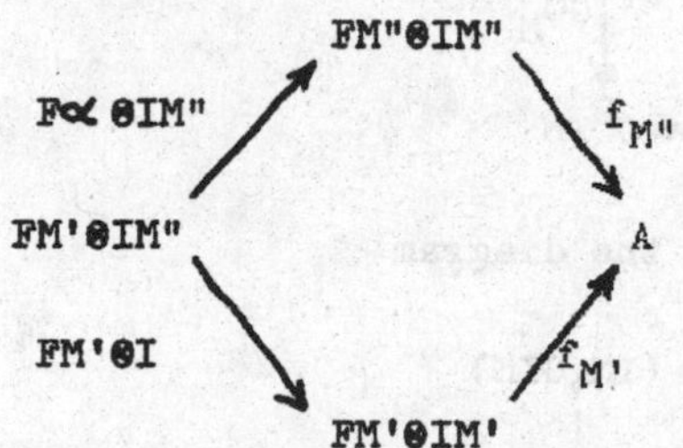

commutes. So if $A \in \underline{A}$, $\varepsilon A\colon rsA \longrightarrow A$ is given by the family of evaluation morphisms

$$(IM,A)\otimes IM \longrightarrow A$$

for $M \in \underline{M}$. If $F \in (\underline{M}^*,A)$ and $M \in \underline{M}$, then

$$\eta F(M)\colon FM \longrightarrow (IM,rF)$$

is the morphism corresponding under hom-$\otimes$ adjointness to j_M.

If $M \in \underline{M}$, then

$$\eta\, sIM(M)\colon (IM,IM) \longrightarrow (IM,GIM),$$

and we put

$$\Theta_M = U(\eta\, sIM(M))(IM) \quad .$$

The naturality of η shows quickly that Θ_M is natural in M. Furthermore, Θ_M is a coalgebra structure for IM. Indeed, from the relation (ε, η): r ⊣ s one sees immediately that $\varepsilon IM \cdot \Theta_M = IM$, so it remains to show that

$$\begin{array}{ccc} IM & \xrightarrow{\Theta_M} & GIM \\ {\scriptstyle \Theta_M}\downarrow & & \downarrow{\scriptstyle G\Theta_M} \\ GIM & \xrightarrow[\delta IM]{} & G^2IM \end{array}$$

commutes. Since η is natural, the diagram

$$\begin{array}{ccc} (IM,IM) & \xrightarrow{\eta\, sIM(M)} & (IM,GIM) \\ {\scriptstyle \eta\, sIM(M)}\downarrow & & \downarrow{\scriptstyle \eta\, srsIM(M)} \\ (IM,GIM) & \xrightarrow[sr\eta\, sIM(M)]{} & (IM,G^2IM) \end{array}$$

commutes. Applying U and chasing the identity we find that

$$\delta IM \cdot \Theta_M = U(\eta\, srsIM(M))(\Theta_M).$$

Now we claim that for any $A \in \underline{A}$,

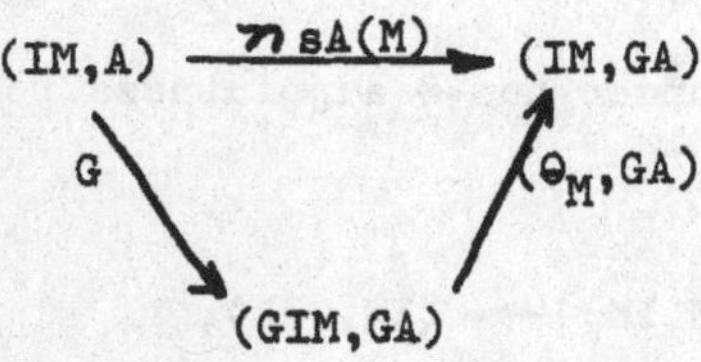

commutes. It is enough to check this after application of U, so let $f\colon IM \longrightarrow A$ be an arbitrary $\underline{A}$-morphism. Then $sf\colon sIM \longrightarrow sA$ is a morphism of functors, and since η is natural

$$\begin{array}{ccc} (IM,IM) & \xrightarrow{\ \eta\, sIM(M)\ } & (IM,GIM) \\ {\scriptstyle (IM,f)}\Big\downarrow & & \Big\downarrow{\scriptstyle (IM,Gf)} \\ (IM,A) & \xrightarrow{\ \eta\, sA(M)\ } & (IM,GA) \end{array}$$

commutes. Chasing the identity, after application of U, gives the result. Applying this to the case $A = GIM$ we obtain

$$U(\eta\, srsIM(M))(\theta_M) = G\theta_M \circ \theta_M,$$

and thus θ_M is a coalgebra structure for IM which is natural in M. Hence we have a lifting

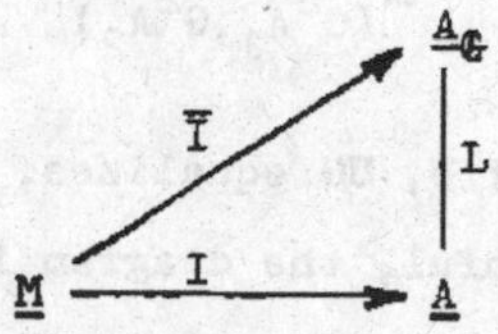

defined by

$$\bar{I}M = (IM, \theta_M)$$
$$\bar{I}\alpha = I\alpha .$$

We can give $\underline{A}_{\mathfrak{G}}$ an $\underline{A}$-structure by requiring, for each (A,θ) and (B,ξ) in $\underline{A}_{\mathfrak{G}}$, that the diagram

$$\begin{array}{ccccc} ((A,\theta),(B,\xi)) & \longrightarrow & (A,B) & \xrightarrow{(A,\xi)} & (A,GB) \\ & & {\scriptstyle G}\searrow & & \nearrow{\scriptstyle (\theta,GB)} \\ & & & (GA,GB) & \end{array}$$

be an equalizer in $\underline{A}$. The morphism

$$((A,\theta),(B,\xi)) \longrightarrow (A,B)$$

is nothing but a lifting of L, so L becomes a strong functor for this A-structure on $\underline{A}_{\mathbf{G}}$. Also, the functor $R: \underline{A} \longrightarrow \underline{A}_{\mathbf{G}}$ given by $RA = (GA, \delta A)$ is strong, as can be seen from the diagram

$$\begin{array}{ccccc} & {\scriptstyle R} \swarrow\!\!\cdots & (A_1,A_2) & & \\ & & \downarrow{\scriptstyle G} & & \\ ((GA_1,\delta A_1),(GA_2,\delta A_2)) & \longrightarrow & (GA_1,GA_2) & \xrightarrow{(GA_1,\delta A_2)} & (GA_1,G^2A_2) \\ & & {\scriptstyle G}\searrow & & \nearrow{\scriptstyle (\delta A_1,G^2A_2)} \\ & & & (G^2A_1,G^2A_2) & \end{array}$$

The fact that after application of U, UG equalizes, says simply that δ is natural. U being faithful, the diagram is equalized at the $\underline{A}$-level and we obtain the indicated strengthening of R. An easy argument using the strength of R and L and the faithfulness of U shows that the adjointness

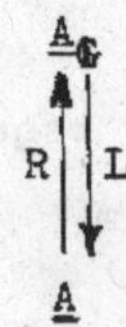

is strong.

Since $\underline{A}_{\mathcal{G}}$ is an $\underline{A}$-category and I lifts to $\overline{I}$, we can define

$$\overline{s}: \underline{A}_{\mathcal{G}} \longrightarrow (M^*, \underline{A})$$

by

$$\overline{s}(A,\Theta)(M) = (\overline{I}M, (A,\Theta)).$$

Now we have an obvious natural transformation

$$j: \overline{s}(A,\Theta) \longrightarrow sA,$$

and we claim that for each (A,Θ),

$$s(A,\Theta) \xrightarrow{j} sA \overset{s\Theta}{\underset{\eta sA}{\rightrightarrows}} srsA$$

is an equalizer. Well, this will be true iff for each $M \in \underline{M}$,

$$((IM,\Theta_M),(A,\Theta)) \longrightarrow (IM,A) \overset{(IM,\Theta)}{\underset{\eta sA(M)}{\rightrightarrows}} (IM,GA)$$

is an equalizer in $\underline{A}$. But by definition,

$$\begin{array}{ccccc} ((IM,\Theta_M),(A,\Theta)) & \longrightarrow & (IM,A) & \xrightarrow{(IM,\Theta)} & (IM,GA) \\ & & {\scriptstyle G}\searrow & & \nearrow{\scriptstyle (\Theta_M,GA)} \\ & & & (GIM,\ GA) & \end{array}$$

is an equalizer and we know that

$$\eta sA(M) = (\Theta_M, GA) \circ G,$$

so we are done.

The Eilenberg-Moore comparison

$$\bar{r}: (M^*,\underline{A}) \longrightarrow A_G$$

is given by

$$\bar{r}F = (rF, r\eta F),$$

and we shall show that $\bar{r}$ is strongly coadjoint to $\bar{s}$, but before doing this we need a remark. Namely, $F \in (\underline{M}^*,\underline{A})$ defines a functor

$$(F,-): (\underline{M}^*,\underline{A}) \longrightarrow \underline{A}$$

If we give

$$\mathcal{F}: \underline{A} \longrightarrow (\underline{M}^*,\underline{A})$$

by

$$\mathcal{F}A(M) = A \otimes FM$$

for $A \in \underline{A}$ and $M \in \underline{M}$, then we claim $\mathcal{F}$ is strongly coadjoint to $(F,-)$. To see this, let $G \in (\underline{M}^*,\underline{A})$. Then $(\mathcal{F}A,G)$ is the equalizer of the pair

$$\prod_{M \in \underline{M}} (\mathcal{F}A(M),GM) \rightrightarrows \prod_{\alpha:M'' \rightarrow M'} (\mathcal{F}A(M'),GM'')$$

or, equivalently, of the pair

$$\prod_{M \in \underline{M}} (A,(FM,GM)) \rightrightarrows \prod_{\alpha:M'' \rightarrow M'} (A,(FM',GM''))$$

However, $(A,-)$ preserves limits so this is the equalizer of the pair

$$(A,\prod_{M \in \underline{M}} (FM,GM)) \rightrightarrows (A,\prod_{\alpha:M'' \rightarrow M'} (FM',GM''))$$

which, for the same reason, is $(A,(F,G))$. In particular, $(F,-)$ preserves limits.

Now, by definition, if $F \in (\underline{M}^*,\underline{A})$ and $(A,\Theta) \in \underline{A}_{\mathbb{G}}$, then $((rF, r\eta F),(A,\Theta))$ is the equalizer of the pair

$$\begin{array}{ccc} (rF,A) & \xrightarrow{(rF,\Theta)} & (rF,GA) \\ & {\scriptstyle G}\searrow \qquad \nearrow{\scriptstyle (r\eta F,GA)} & \\ & (GrF,GA) & \end{array}$$

Denoting, for a moment, the strong adjointness isomorphism by

$$a\colon (rF,A) \longrightarrow (F,sA),$$

the above pair corresponds to the pair

$$\begin{array}{ccc} (F,sA) & \xrightarrow{(F,s\Theta)} & (F,srsA) \\ & \searrow \qquad \nearrow{\scriptstyle (\eta F,srsA)} & \\ & (srF,srsA) & \end{array}$$

where the unnamed morphism is the composite

$$(F,sA) \xrightarrow{a^{-1}} (rF,A) \xrightarrow{rs} (rsrF,rsA) \xrightarrow{a} (srF,srsA).$$

More explicitly, this is the composite

$$(F,sA) \xrightarrow{r} (rF,rsA) \xrightarrow{(rF,\varepsilon A)} (rF,A) \xrightarrow{rs} (rsrF,rsA)$$

$$\xrightarrow{s} (srsrF,srsA) \xrightarrow{(\eta srF,srsA)} (srF,srsA)$$

Thus, on the set level, starting with $\gamma\colon F \longrightarrow sA$ we obtain the composite

$$\mathrm{srF} \xrightarrow{\eta\,\mathrm{srF}} \mathrm{srsrF} \xrightarrow{\mathrm{srsr}\Upsilon} \mathrm{srsrsA} \xrightarrow{\mathrm{srs}\varepsilon\mathrm{A}} \mathrm{srsA}.$$

Composing further with ηF yields the commutative diagram

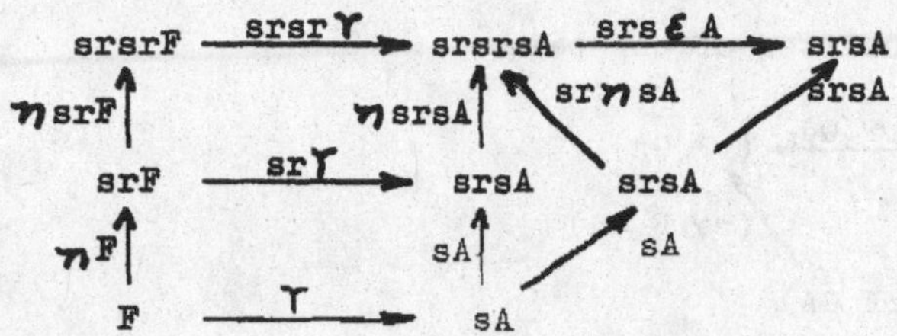

Thus, since U is faithful we have shown that the composite

$$(\mathrm{F},\mathrm{sA}) \longrightarrow (\mathrm{srF},\mathrm{srsA}) \xrightarrow{(\eta\mathrm{F},\mathrm{srsA})} (\mathrm{F},\mathrm{srsA})$$

is nothing but (F,ηsA). Hence, ((rF,rηF), (A,Θ)) can be identified with the equalizer of the pair

$$(\mathrm{F},\mathrm{sA}) \underset{(\mathrm{F},\eta\mathrm{sA})}{\overset{(\mathrm{F},\mathrm{s}\Theta)}{\rightrightarrows}} (\mathrm{F},\mathrm{srsA}) ,$$

but by previous remarks this is (F, $\bar{s}$(A,Θ)), so that $\bar{r}$ is strongly coadjoint to $\bar{s}$. Again, this implies that $\bar{r}$ and $\bar{s}$ are both strong.

From this point on, the equivalence theorem goes as in [1]. We shall sketch the result, both for the convenience of the reader, and because we need a corollary of the proof itself in the first example. So, let

$$\bar{\varepsilon}: \bar{r}\bar{s} \longrightarrow \underline{A}_{\mathbb{G}}$$

$$\bar{\eta}: (\underline{M}^*,\underline{A}) \longrightarrow \bar{s}\bar{r}$$

denote the counit and unit respectively of the strong adjointness $\bar{r} \dashv \bar{s}$ (we will indicate in a moment how these arise).

Theorem.

(a) If $(A,\Theta) \in \underline{A}_G$, then $\overline{\varepsilon}(A,\Theta)\colon \overline{r}\,\overline{s}(A,\Theta) \longrightarrow (A,\Theta)$ is an isomorphism iff

$$r\overline{s}(A,\Theta) \xrightarrow{rj} rsA \underset{r\eta sA}{\overset{rs\Theta}{\rightrightarrows}} (rs)^2A$$

is an equalizer in $\underline{A}$.

(b) If $\overline{\varepsilon}(A,\Theta)\colon \overline{r}\,\overline{s}(A,\Theta) \longrightarrow (A,\Theta)$ is an isomorphism for each $(A,\Theta) \in \underline{A}_G$, then $\overline{\eta}F\colon F \longrightarrow \overline{s}\,\overline{r}F$ is an isomorphism for each $F \in (M^*,A)$ iff r reflects isomorphisms.

__Proof__: From the definitions, it is immediate that $(A,\Theta) \in \underline{A}_G$ iff

$$A \xrightarrow{\Theta} rsA \underset{r\eta sA}{\overset{rs\Theta}{\rightrightarrows}} (rs)^2A$$

is an equalizer in $\underline{A}$. Since rj equalizes $rs\Theta$ and $r\eta sA$, there is a unique morphism $r\overline{s}(A,\Theta) \longrightarrow A$ making

$$\begin{array}{ccccc} r\overline{s}(A,\Theta) & \xrightarrow{rj} & rsA & \underset{r\eta sA}{\overset{rs\Theta}{\rightrightarrows}} & (rs)^2A \\ & \searrow & \uparrow{\scriptstyle\Theta} & & \\ & & A & & \end{array}$$

commute, and it is not hard to see that this morphism is precisely $L(\overline{\varepsilon}(A,\Theta))$. Since L reflects isomorphisms, $\overline{\varepsilon}(A,\Theta)$ is an isomorphism iff $L(\overline{\varepsilon}(A,\Theta))$ is, and this completes the proof of (a).

Let $F \in (\underline{M}^*,\underline{A})$, and assume $\overline{\varepsilon}(A,\Theta)$ is an isomorphism for all $(A,\Theta) \in A_G$. We have seen that

$$\overline{s}\overline{r}F \xrightarrow{j} srF \underset{\eta\, srF}{\overset{sr\eta F}{\rightrightarrows}} (sr)^2F$$

is an equalizer, and since $\eta F\colon F \to srF$ equalizes $sr\eta F$ and ηsrF by naturality, there is a unique natural transformation $F \to \bar{s}\bar{r}F$ making

$$\begin{array}{ccc} F & \xrightarrow{\eta F} & srF \underset{\eta srF}{\overset{sr\eta F}{\rightrightarrows}} (sr)^2F \\ & \searrow \quad \nearrow_{j} & \\ & \bar{s}\bar{r}F & \end{array}$$

commute. This, of course, is $\bar{\eta}F$. By (a), both ηF and j become equalizers upon application of r, making $r\bar{\eta}F$ an isomorphism. Thus $\bar{\eta}F$ is an isomorphism if r reflects isomorphisms. The converse is trivial since L reflects isomorphisms.

Note that if r reflects equalizers, then it follows that ηF is the equalizer of $sr\eta F$ and ηsrF, which is also necessary and sufficient for $\bar{\eta}F$ to be an isomorphism. Combining this with (a), we have the following sufficient condition for equivalence. (This is the dual of Beck's CTT.)

Corollary.

If r preserves and reflects equalizers, then

$$\underline{A}_{G} \underset{\bar{r}}{\overset{\bar{s}}{\rightleftarrows}} (\underline{M}^*, \underline{A})$$

is a strong adjoint equivalence of categories.

2. Examples

(i) Chain complexes and FD-modules.

Let $\underline{\Delta}$ be the simplicial category, i.e., the objects of $\underline{\Delta}$ are sequences $[n] = (0, \ldots, n)$ and a morphism $\alpha\colon [m] \to [n]$

is a monotone function. K will denote a commutative ring with unit, and we will write Mod(K) for the autonomous category of K-modules. Define

$$I: \underline{\underline{\Delta}} \longrightarrow \underline{\text{Mod}}(K)$$

by

$$I[n] = \text{free K-module on the injections } \varepsilon : [q] \longrightarrow [n],$$

and if $\alpha: [m] \longrightarrow [n]$ then

$$I\alpha(\varepsilon) = \begin{cases} \alpha \cdot \varepsilon & \text{if this is injective} \\ 0 & \text{otherwise} \end{cases}$$

One can verify easily that I is a functor. In [1] we analyzed this example without taking into account the autonomous structure of Mod(K), and the reader might find it interesting to compare the two treatments.

From §1 we have a diagram

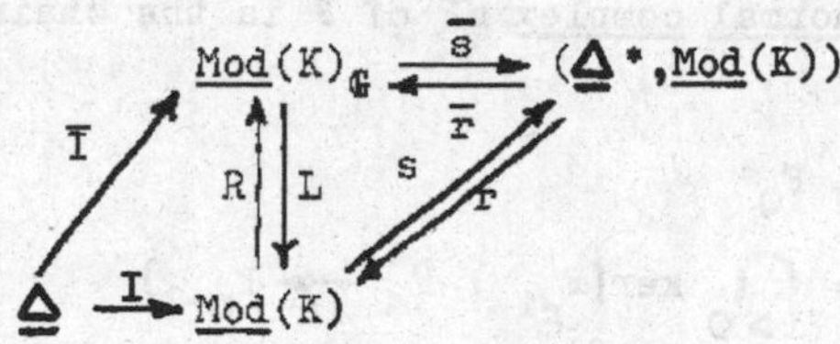

and each adjoint pair is strong. $(\underline{\underline{\Delta}}^*, \underline{\text{Mod}}(K))$ is the category of FD-modules, and if F is an FD-module, then rF is defined by requiring that

$$\coprod_{\alpha:[r]\to[s]} F_s \otimes I[r] \underset{\tau_2}{\overset{\tau_1}{\rightrightarrows}} \coprod_{[n]\in\underline{\Delta}} F_n \otimes I[n] \xrightarrow{\rho} rF$$

be a coequalizer in $\underline{\text{Mod}}(K)$ where

$$\tau_1 \cdot i_\alpha = i_{[r]} \cdot F_\alpha \otimes I[r]$$

$$\tau_2 \cdot i_\alpha = i_{[s]} \cdot F_s \otimes I\alpha$$

(Note that, as is customary in the literature, we write F_n for $F[n]$ and F_α for $F\alpha$.)

Let us denote by

$$\varepsilon^i: [n-1] \longrightarrow [n]$$

the injection in $\underline{\Delta}$ that misses i, and by

$$\eta^j: [n+1] \longrightarrow [n]$$

the surjection that covers j twice. (In general, we will write ε for injections and η for surjections.) Recall from [4] that if $F \in (\underline{\Delta}^*, \underline{\text{Mod}}(K))$ the <u>normal</u> <u>complex</u> NF of F is the chain complex given by

$$(NF)_0 = F_0$$

$$\text{and } (NF)_n = \bigcap_{i>0} \ker\{F_{\varepsilon^i} : F_n \longrightarrow F_{n-1}\}$$

for $n > 0$, with boundary operator induced by F_{ε^0}. For $n \geq 0$, denote by $(DF)_n$ the submodule of F_n generated by all elements of the form $F_{\eta^i}(x)$ for $0 \leq i \leq n-1$ and $x \in F_{n-1}$. If we make F into a chain complex by setting $\partial_n : F_n \to F_{n-1}$ equal to $\sum_{i=1}^{n} (-1)^i F_{\varepsilon^i}$, then DF is a subcomplex of F, and it is well known (see, for example, [4], Satz 3.22) that there is a natural epimorphism of chain complexes

$$\pi : F \to NF$$

which is split by the inclusion, and whose kernel is DF.

In order to identify rF, define

$$\phi_n : F_n \otimes I[n] \to \coprod_m (NF)_m$$

by

$$\phi_n : (x \otimes \varepsilon) = i_q \pi_q (F_\varepsilon x)$$

where $\varepsilon : [q] \to [n]$ and i_q denotes the canonical injection

$$(NF)_q \to \coprod_m (NF)_m$$

A routine verification from the definitions shows that the ϕ_n define a map ϕ such that

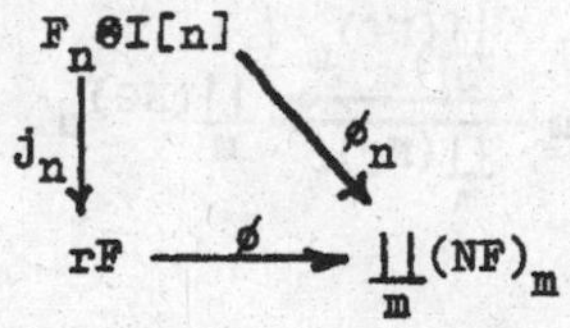

commutes, where $j_n = \rho \cdot i_{[n]}$. It is easily seen that the map

$$F_n \to rF$$

given by sending x to $j_n(x \otimes 1_{[n]})$ vanishes on $(DF)_n$, and hence defines

$$\phi'_n : (NF)_n \longrightarrow rF,$$

and thus

$$\phi' : \coprod_m (NF)_m \longrightarrow rF .$$

The straightforward calculation that $\phi\phi'$ and $\phi'\phi$ are the respective identities will be left to the reader. Since ϕ and ϕ' are obviously natural, we may henceforth use $\coprod_m (NF)_m$ for rF.

The fact that r preserves equalizers is essentially trivial. In fact, a diagram

$$E \xrightarrow{e} F \underset{g}{\overset{f}{\rightrightarrows}} G$$

of FD-modules is an equalizer iff for each $n \geq 0$

$$E_n \xrightarrow{e_n} F_n \underset{g_n}{\overset{f_n}{\rightrightarrows}} G_n$$

is, and

$$\coprod_m (NE)_m \xrightarrow{\coprod_m (Ne)_m} \coprod_m (NF)_m \underset{\coprod_m (Ng)_m}{\overset{\coprod_m (Nf)_m}{\rightrightarrows}} \coprod_m (NG)_m$$

is an equalizer iff each

$$(NE)_n \xrightarrow{(Ne)_n} (NF)_n \underset{(Ng)_n}{\overset{(Nf)_n}{\rightrightarrows}} (NG)_n$$

is. Now if we consider the diagram

$$\begin{array}{ccccc} (NE)_n & \xrightarrow{(Ne)_n} & (NF)_n & \overset{(Nf)_n}{\underset{(Ng)_n}{\rightrightarrows}} & (NG)_n \\ \downarrow i_E & & \downarrow i_F & & \downarrow i_G \\ E_n & \longrightarrow & F_n & \overset{f_n}{\underset{g_n}{\rightrightarrows}} & G_n \end{array}$$

where the i's are the natural inclusions, then it follows immediately from the naturality of e and the fact that each i is monic, that if the bottom row is an equalizer so is the top.

The assertion that r reflects isomorphisms is precisely Lemma 3.17 of [4], so

$$\underline{\text{Mod}}(K)_{\mathbb{G}} \underset{\bar{r}}{\overset{\bar{s}}{\rightleftarrows}} (\underline{\Delta}^*, \underline{\text{Mod}}(K))$$

is a strong adjoint equivalence of categories by the equivalence theorem of §1.

To complete the result, we will use cotriple methods to identify the category $\underline{\text{Mod}}(K)_{\mathbb{G}}$. First, consider the cotriple $\mathbb{G}$. If $M \in \underline{\text{Mod}}(K)$, then

$$GM = \coprod_n (NsM)_n$$

where

$$(NsM)_n = \{\phi\colon I[n] \longrightarrow M \mid \phi \cdot I\varepsilon^i = 0 \quad i > 0\} \quad .$$

The image of $I\varepsilon^i\colon I[n-1] \longrightarrow I[n]$ is the free submodule of I[n] with basis those injections $\varepsilon\colon [q] \longrightarrow [n]$ that factor through ε^i, and these are precisely the injections that miss i. If $\phi\colon I[n] \longrightarrow M$ vanishes on the image of $I\varepsilon^i$ for all $i > 0$, then

$\phi(\varepsilon) = 0$ for each ε: $[q] \to [n]$ that misses any one of the places 1, ..., n, and conversely. Thus $(NSM)_n$ can be identified with the K-module

$$(I'[n],M) ,$$

where I'[n] is the free K-module on $I_{[n]}$ and ε^0. This is just $M \oplus M$ if $n > 0$ and M is $n = 0$. $(\varepsilon M)_n$ is given by evaluation on $1_{[n]}$, or, equivalently, projection on the first factor. Now, with this analysis we can derive a formula for δM: $GM \to G^2M$ and show directly that $\underline{Mod}(K)_{\mathfrak{G}}$ is the category of chain complexes over K, but the following seems a somewhat more agreeable approach. Namely, let C(K) denote the category of positive chain complexes over K. Motivated by the above, let

$$s': \underline{Mod}(K) \to C(K)$$

be defined by

$$(s'M)_n = \begin{cases} M \oplus M & \text{if } n > 0 \\ M & \text{if } n = 0 \end{cases} ,$$

with ∂_n: $(s'M)_n \to (s'M)_{n-1}$ given by

$$\partial_n(m_1,m_2) = (m_2,0) \quad \text{if } n > 1$$

and

$$\partial_1(m_1,m_2) = m_2 .$$

As indicated above, s' can easily be made to be model induced as in §1. However, since the functors involved here are so simple, and since the functor category obtained in this manner must be restricted anyway to give chain complexes, it did not seem worthwhile to do this.

Let us examine the chain maps between an arbitrary chain complex $C = (C_n)$ and s'M. Well, such a map ϕ is given by a family

$$\phi_n \colon C_n \longrightarrow (sM)_n \quad n \geq 0$$

such that $\partial_n \phi_n = \phi_{n-1} \partial_n$. For $n = 0$, ϕ_0 is a single map

$$\phi_0 = f_0 \colon C_0 \longrightarrow M,$$

and for $n > 0$, ϕ_n is given by a pair

$$\phi_n = \langle f_n, g_n \rangle \colon C_n \longrightarrow M \oplus M .$$

We must have $\partial_1 \phi_1 = \phi_0 \partial_1$, i.e.,

$$\partial_1 \langle f_1, g_1 \rangle = f_0 \partial_1 ,$$

so

$$g_1 = f_0 \partial_1$$

and f_1 is arbitrary. Proceeding by induction, we assume

$$\phi_{n-1} = \langle f_{n-1}, f_{n-2} \partial_{n-1} \rangle .$$

Then

$$\partial_n \langle f_n, g_n \rangle = \langle f_{n-1}, f_{n-2} \partial_{n-1} \rangle \partial_n$$

gives

$$\langle g_n, 0 \rangle = \langle f_{n-1} \partial_n, 0 \rangle .$$

Hence again f_n is arbitrary and $g_n = f_{n-1} \partial_n$. Thus we see that a chain map ϕ is determined by an arbitrary family $f_n \colon C_n \longrightarrow M$, or, equivalently, by a single map

$$f \colon \coprod_m C_m \longrightarrow M$$

Therefore, if we define

$$r' \colon C(K) \longrightarrow \underline{Mod}(K)$$

by $r'C = \coprod_m C_m$, we have

$$r' \dashv s' ,$$

and the adjointness is obviously strong.

The point of all this is that if $\mathbb{G}'$ is the cotriple induced by $r' \dashv s'$, then $\mathbb{G}'$ is naturally equivalent to $\mathbb{G}$ as a cotriple on $\underline{Mod}(K)$. (We have already verified that the functor parts are equivalent, and the compatibility of this equivalence with the counits is evident. Checking the relevant assertion concerning the comultiplication requires a little patience due to the numerous identifications we have made.) As a result, the categories $\underline{Mod}(K)_{\mathbb{G}}$ and $\underline{Mod}(K)_{\mathbb{G}'}$ are equivalent (in a particularly simple way), and it only remains to show that $\underline{Mod}(K)_{\mathbb{G}'}$ is equivalent to $C(K)$.

Let

$$\eta' \colon C(K) \longrightarrow s'r'$$

and

$$\varepsilon' \colon r's' \longrightarrow \underline{Mod}(K)$$

denote, respectively, the unit and counit of the adjointness $r' \dashv s'$. Then the Eilenberg-Moore comparison

$$\bar{r}': C(K) \longrightarrow \underline{\text{Mod}}(K)_{\mathbb{G}'}$$

is given by $\bar{r}'C = (r'C, r'\eta'C)$, and $\bar{r}'$ has a strong adjoint

$$\bar{s}': \underline{\text{Mod}}(K)_{\mathbb{G}'} \longrightarrow C(K)$$

given by requiring, for $(M,\theta') \in \underline{\text{Mod}}(K)_{\mathbb{G}'}$, that

$$\bar{s}'(M,\theta') \longrightarrow s'M \underset{\eta's'M}{\overset{s'\theta'}{\rightrightarrows}} s'r's'M$$

be an equalizer in C(K). Since r' obviously preserves and reflects equalizers, we may use the corollary to the equivalence theorem of §1 to assert that

$$\underline{\text{Mod}}_{\mathbb{G}'} \underset{\bar{r}'}{\overset{\bar{s}'}{\rightleftarrows}} C(K)$$

is a strong adjoint equivalence of categories. Thus we recover the theorem of Dold and Kan ([3] and [7] respectively) concerning the equivalence of the categories of FD-modules and chain complexes. A simple modification extends the result to an arbitrary abelian category.

(ii) Torsion groups

Let $\underline{\text{Ab}}$ denote the category of abelian groups, and write $\underline{C}$ for the full subcategory of $\underline{\text{Ab}}$ whose objects are cyclic groups $Z_n = Z/nZ$ where Z is the group of integers and $n > 0$. Let

$$I: \underline{C} \longrightarrow \underline{\text{Ab}}$$

be the inclusion, which we will henceforth supress from the notation.

If $A \in \underline{Ab}$ and $n > 0$, let T_nA denote the kernel of the homomorphism $n: A \longrightarrow A$ given by multiplication by n. TA will denote the torsion subgroup of A, which consists of the elements of A of any finite order. Clearly,

$$(Z_n, A) \xrightarrow[\approx]{} T_nA$$

where the isomorphism is given by $\phi \rightsquigarrow \phi(1)$. From this it follows that

$$(Z_n, Z_m) = Z_d,$$

where $d = \gcd(m,n)$. Furthermore, a choice of generator is given by the homomorphism $1 \rightsquigarrow \frac{m}{d}$.

From §1, I induces a strong adjoint pair

$$\underline{Ab} \underset{r}{\overset{s}{\rightleftarrows}} (\underline{C}^*, \underline{Ab})$$

and what we need now is a convenient description of rF -- at least if $F \in (\underline{C}^*, \underline{Ab})$ is additive. To this end, let $\underline{C}_0$ be the subcategory of $\underline{C}$ with the same objects but where

$$\alpha: Z_n \longrightarrow Z_m$$

in $\underline{C}_0$ iff $m|n$, in which case α is the epimorphism $1 \rightsquigarrow 1$. Let $J: \underline{C}_0 \longrightarrow \underline{C}$ be the inclusion, and note that $\underline{C}_0^*$ is a directed set. Now let $F: \underline{C}^* \longrightarrow \underline{Ab}$ be additive and $A \in \underline{Ab}$. Then a homomorphism

$$\phi\colon \varinjlim(F \circ J^*) \longrightarrow A$$

consists of a sequence of homomorphisms

$$\phi_n\colon F(Z_n) \longrightarrow A$$

such that for each of the above α's,

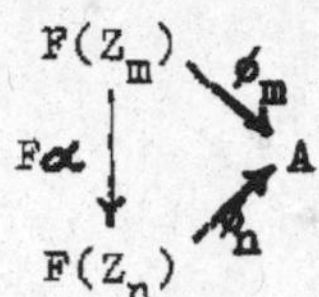

commutes. A morphism

$$(F,sA) \longrightarrow (\varinjlim(F\circ J^*),A)$$

is given by sending

$$\Upsilon \rightsquigarrow (\Upsilon'_n) ,$$

where $\Upsilon'_n\colon F(Z_n) \longrightarrow A$ is defined by $\Upsilon'_n(x) = \Upsilon_n(x)(1)$. Suppose $\phi\colon \varinjlim(F\circ J^*) \longrightarrow A$, and define

$$\phi'_n\colon F(Z_n) \longrightarrow (Z_n,A)$$

by $\phi'_n(x)(1) = \phi_n(x)$. This makes sense because since F is additive, $n\cdot F(Z_n) = 0$ so $n\cdot \mathrm{im}_{\phi_n} = 0$ and $\mathrm{im}\ \phi_n \subset T_nA$. Let $\beta\colon Z_n \longrightarrow Z_m$ be an arbitrary morphism in $\underline{C}$. To show that the ϕ'_n define a natural transformation $\phi'\colon F \longrightarrow sA$, consider the diagram

$$\begin{array}{ccc} F(Z_m) & \xrightarrow{\phi'_m} & (Z_n,A) \\ {\scriptstyle F\beta}\downarrow & & \downarrow{\scriptstyle (\beta,A)} \\ F(Z_n) & \xrightarrow{\phi'_n} & (Z_n,A) \end{array}$$

Since F and sA are additive it suffices to show this commutes when β is the generator of (Z_n,Z_m). In that case β factors as

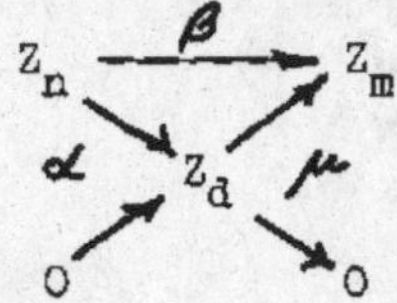

where $\alpha(1) = 1$ and $\mu(1) = \frac{m}{d}$ for $d = \gcd(m,n)$. Thus the diagram becomes

$$\begin{array}{ccc} F(Z_m) & \xrightarrow{\phi'_m} & (Z_m,A) \\ {\scriptstyle F\mu}\downarrow & & \downarrow{\scriptstyle (\mu,A)} \\ F(Z_d) & \xrightarrow{\phi'_d} & (Zd,A) \\ {\scriptstyle F\alpha}\downarrow & & \downarrow{\scriptstyle (\alpha,A)} \\ F(Z_n) & \xrightarrow{\phi'_n} & (Z_n,A) \end{array}$$

The bottom diagram commutes by assumption, so we are left with the top. For this, consider the composite

$$Z_m \xrightarrow{\alpha'} Z_d \xrightarrow{\mu} Z_m$$

where $\alpha'(1) = 1$. This is $\frac{m}{d}$ times the identity, so the diagram

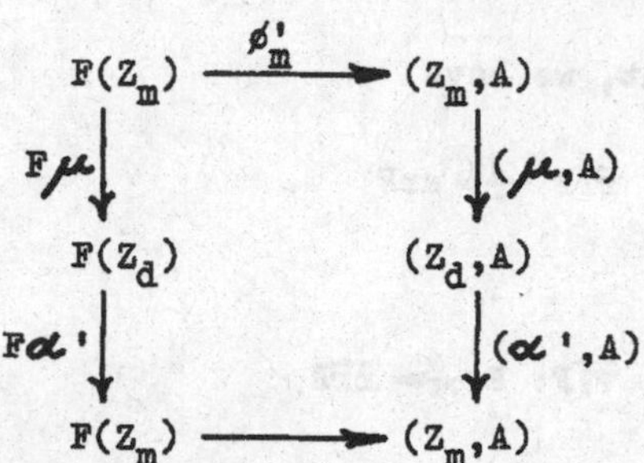

commutes. Inserting ϕ'_d in the middle makes the resulting bottom square commutative by hypothesis. Since (α', A) is monic, the top will also commute. Thus ϕ' is a natural transformation, and

$$(F, sA) \longrightarrow (\underrightarrow{\lim}(F \circ J^*), A)$$

is an isomorphism, which gives

$$rF = \underrightarrow{\lim}(F \circ J^*)$$

If $F = sA$ for $A \in \underline{Ab}$, then it is clear that

$$GA = rsA \approx \underrightarrow{\lim}\,(sA \circ J^*) = TA$$

$\varepsilon A\colon GA \longrightarrow A$ becomes the inclusion $TA \longrightarrow A$ under this identification. Thus the cotriple $\mathbf{G}$ is idempotent (meaning $\delta A\colon GA \xrightarrow{\approx} G^2 A$ for $A \in \underline{Ab}$) and $\underline{Ab}_{\mathbf{G}}$ is the full subcategory of $\underline{Ab}$ consisting of abelian groups A for which $\varepsilon A\colon GA \xrightarrow{\approx} A$, i.e., the full subcategory of torsion groups. It is shown in [1] that idempotence makes

$$\overline{\varepsilon}\colon \overline{r}\overline{s} \longrightarrow \underline{Ab}_{\mathbf{G}}$$

a natural equivalence.

Since $\mathfrak{G}$ is idempotent, we have

$$\overline{s}\overline{r}F \xrightarrow[\approx]{i} srF$$

for $F \in (\underline{C}^*,\underline{Ab})$, so

$$\overline{\eta}F: F \xrightarrow{\approx} \overline{s}\overline{r}F$$

iff

$$\eta F: F \xrightarrow{\approx} srF .$$

If F is additive, it is easy to see that under the identification

$$rF \approx \underrightarrow{\lim} (F \circ J^*) ,$$

$$\eta F(Z_n): F(Z_n) \longrightarrow (Z_n, rF)$$

becomes the factorization of the canonical injection

$$i_n: F(Z_n) \longrightarrow \underrightarrow{\lim} (F \circ J^*)$$

through the inclusion

$$T_n(rF) \longrightarrow rF$$

Call F <u>left</u> <u>exact</u> if for each exact sequence

$$Z_r \longrightarrow Z_s \longrightarrow Z_q \longrightarrow 0$$

in $\underline{C}$,

$$0 \longrightarrow F(Z_q) \longrightarrow F(Z_s) \longrightarrow F(Z_n)$$

is exact in $\underline{Ab}$. Left exactness is clearly a necessary condition for

$$\eta F : F \xrightarrow{\sim} srF$$

since srF has this property. On the other hand, let F be left exact. Then for every $\alpha : Z_n \longrightarrow Z_m$ in $\underline{C}_0$,

$$F\alpha : F(Z_m) \longrightarrow F(Z_n)$$

is monic and hence by directedness the same is true of each i_n, and thus of each

$$\eta F(Z_n) : F(Z_n) \longrightarrow (Z_n, rF),$$

so that ηF is monic. Now suppose $y \in T_n(rF)$. Then for some m there is an $x \in F(Z_m)$ such that $i_m(x) = y$. We may assume $n|m$, otherwise let $m' = nm$. Then we have $\alpha : Z_{m'} \longrightarrow Z_m$ in $\underline{C}_0$ and thus

$$\begin{array}{ccc} F(Z_m) & \xrightarrow{F\alpha} & F(Z_{m'}) \\ & {\scriptstyle i_m}\searrow \quad \swarrow{\scriptstyle i_{m'}} & \\ & \varinjlim (F \circ J^*) & \end{array}$$

so $F\alpha(x)$ will serve our purpose as well as x. Now

$$Z_m \xrightarrow{n} Z_m \xrightarrow{\alpha} Z_n \longrightarrow 0$$

is exact, so in the following diagram the top row is exact:

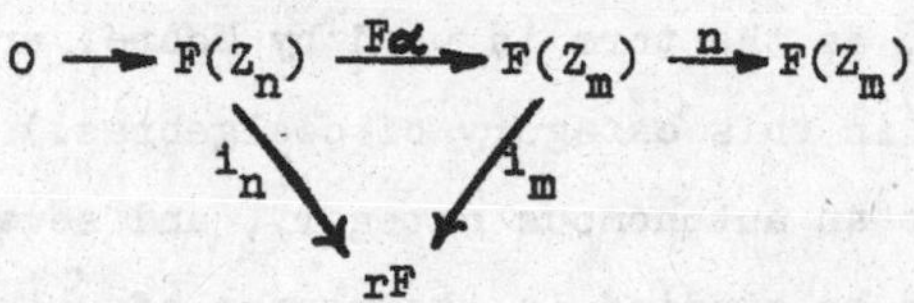

(In each case, n denotes multiplication by n, i.e., n times the identity.) Now, since $i_m(x) = y$ and $ny = 0$ we have $nx = 0$ since i_m is monic. Thus $x = F\alpha(x')$ and

$$y = i_n(x') ,$$

making each $\eta F(Z_n)$ epic. As a result,

$$\eta F: F \xrightarrow{\approx} srF$$

iff F is additive and left exact. (Note that the first condition would follow from the second if C has coproducts which, however, is not the case.)

It should be more or less evident that the model category can be reduced by considering only cyclic groups of prime power order. In this case GA turns out to be the direct sum of the p-primary parts of A which is well-known to be TA. The analysis of rF becomes slightly more complicated under this reduction, however, so it seemed desirable to use the extra models.

(iii) Topological examples

Here we shall understand by Top the category of compactly generated spaces. (For definitions see [1] where these are mistakenly called compactly generated weakly Hausdorff -- actually the weakly Hausdorff spaces, as the term is used by Moore, are exactly the regular objects in this category of coalgebras.) This Top is well-known to be an autonomous category, and several of the examples from [1] can be studied in the manner of §1 simply by letting the model functor take values in compactly generated spaces rather than in all topological spaces. In the interests of brevity, we will only sketch what happens in two of these cases.

First, let

$$I: \underline{\Delta} \longrightarrow \underline{\text{Top}}$$

be the usual functor given by

$$I[n] = \Delta_n \text{ - the standard n-simplex}$$

and

$$I\alpha = \Delta_\alpha \text{ - the unique affine map determined by } \alpha \text{ on the vertices}$$

Then, from §1 we obtain a strong adjoint pair

$$\underline{\text{Top}} \underset{r}{\overset{s}{\rightleftarrows}} (\underline{\Delta}^*, \underline{\text{Top}})$$

If $X \in$ Top, then the underlying set of sX, in each dimension, gives the usual singular complex of X. If K is a simplicial object in Top, i.e., $K \in (\underline{\Delta}^*, \underline{\text{Top}})$, then one verifies easily that the underlying set of rK is the same as the underlying set of $|K'|$, where $|K'|$ is the usual geometric realization of the simplicial set K' obtained by forgetting, in each dimension, the topology of K. The topology of rK is coarser than that of $|K'|$, however. The coalgebras over the strong cotriple do not seem to be known objects.

For another example, let G be a group in Top, and let $\underline{G}$ be the category with one object G whose morphisms are the elements $g \in G$. Define

$$I: \underline{G} \longrightarrow \underline{Top}$$

by $$IG = G$$

and $$Ig = Lg = \text{left translation by } g \in G.$$

Here ($\underline{G}^*$,$\underline{Top}$) can be identified with the category of right G-spaces and equivariant maps, where G is regarded as a discrete group. The coalgebras over the strong model induced cotriple are ordinary right G-spaces and equivariant maps, which are coreflective in, but not equivalent to, ($\underline{G}^*$,$\underline{Top}$).

References

[1] H. Appelgate - M. Tierney, Categories with models, Springer Lecture Notes in Mathematics No. 80, 1969, 156-244.

[2] M. Bunge, Relative functor categories and categories of algebras, J. Algebra 11 (1969), 64-101.

[3] A. Dold, Homology of symmetric products and other functors of complexes, Ann. of Math. 68 (1958), 54-80.

[4] A. Dold - D. Puppe, Homologie nicht additiver Funktoren. Anwendungen, Ann. Inst. Fourier, 11 (1961), 201-312.

[5] S. Eilenberg - G.M. Kelley, Closed categories, Proceedings of the Conference on Categorical Algebra - La Jolla 1965. Springer, Berlin-Heidelberg-New York 1966.

[6] S. Eilenberg - J. Moore, Adjoint functors and triples, Ill. J. Math 9 (1965), 381-398.

[7] D. Kan, Functors involving css complexes, Trans. Amer. Math. Soc. 87 (1958), 330-346.

[8] F.E.J. Linton, Autonomous categories and duality of functors, J. Algebra 2 (1965), 315-349.

ADJUNCTION FOR ENRICHED CATEGORIES

by G. M. Kelly

Received April 30, 1969

1. Introduction. The following pages contain, in very summary form, some observations on adjunction in the context of closed categories. These were originally intended to form part of a sequel to [3], and for various reasons have lain unpublished for some time. To some extent they overlap the recent work of others: Marta Bunge in her thesis [2] considers adjunction for $\underline{V}$-functors where $\underline{V}$ is a closed category; both she and Kock [4] consider the closely related matter of triads (= triples = monads = standard constructions) in the same context; various people including Beck [1] have considered what we call *tensored* $\underline{V}$-categories. To sort out in detail precisely what is original in the following observations would only cause further delays; so we present them as they stand, in the form of a summary exposition.

Further thinking about closed categories since the completion of [3] has led to the conclusion that many important constructions can be carried out only in the presence of a symmetric tensor product. While some of the things below, then, can be stated and proved in a more general setting, we restrict ourselves to what seems to be the main case of interest, and agree that henceforth *closed category* shall mean what in [3] was called "symmetric monoidal closed category".

2. Adjunction in any 2-category. It is clear that one can define

adjunction in any 2-category (= what was called a <u>hypercategory</u> in [3]). We write $\varepsilon, \eta : S \dashv T: \underline{A},\underline{B}$ and we say that S is left-adjoint to T if $T: \underline{A} \longrightarrow \underline{B}$, $S: \underline{B} \longrightarrow \underline{A}$, $\varepsilon: ST \Rightarrow 1$, $\eta: 1 \Rightarrow TS$, with $T\varepsilon \circ \eta T = 1$ and $\varepsilon S \circ S\eta = 1$. It is easy to see that if $S \dashv T: \underline{A},\underline{B}$ and $Q \dashv P: \underline{B},\underline{C}$ then $SQ \dashv PT: \underline{A},\underline{C}$.

2.1. *Let* $\varepsilon, \eta : S \dashv T: \underline{A},\underline{B}$ *and let* $\varepsilon', \eta' : S' \dashv T': \underline{A}',\underline{B}'$. *Let* $P: \underline{A} \longrightarrow \underline{A}'$ *and* $Q: \underline{B} \longrightarrow \underline{B}'$. *Then we have a bijection*

$$(QT,T'P) \cong (S'Q,PS)$$

between 2-*cells* $\lambda: QT \Rightarrow T'P$ *and* 2-*cells* $\mu: S'Q \Rightarrow PS$.

<u>Proof</u>. First take the special case in which $\underline{A} = \underline{B}$, $S = T = 1$, $\varepsilon = \eta = 1$. It is easy to see that a bijection $\lambda \leftrightarrow \mu$ between $(Q,T'P)$ and $(S'Q,P)$ is set up by the equations $\mu = T\lambda \circ \eta Q$, $\lambda = \varepsilon P \circ S\mu$. Duality then gives, in the special case in which $\underline{A}' = \underline{B}'$, $S' = T' = 1$, $\varepsilon' = \eta' = 1$, a bijection $(QT,P) \cong (Q,PS)$. We now get the desired result by combining the two special cases: $(QT,T'P) \cong (S'QT,P) \cong (S'Q,PS)$.

The bijection of 2.1 has certain evident naturality properties. Taking account of these we easily get:

<u>Corollary</u>. *If* $S \dashv T$ *and* $S' \dashv T$ *then* $S' \cong S$.

If Φ is a 2-functor from one 2-category to another, an adjunction $\varepsilon, \eta : S \dashv T: \underline{A},\underline{B}$ in the first 2-category clearly gives rise to an adjunction $\Phi\varepsilon, \Phi\eta : \Phi S \dashv \Phi T: \Phi\underline{A},\Phi\underline{B}$ in the second.

3. <u>Adjunction in the 2-category of V-categories</u>. Let $\underline{V}$ be a closed category. Then $\underline{V}$-categories, $\underline{V}$-functors, and $\underline{V}$-natural

transformations form a 2-category $\underline{V}$-$\underline{Cat}$, adjunction in which will be called $\underline{V}$-adjunction.

3.1. <u>There is a bijection between $\underline{V}$-adjunctions</u> $\varepsilon, \eta: S \dashv T: \underline{A}, \underline{B}$ and $\underline{V}$-<u>natural isomorphisms</u>

$$n = n_{BA}: \underline{A}(SB,A) \longrightarrow \underline{B}(B,TA). \qquad (1)$$

<u>Proof</u>. By the representation theorem for $\underline{V}$-categories ([3] , page 469 Theorem 10.8 and page 548 Proposition 7.9), there is a bijection between $\underline{V}$-natural transformations (not isomorphisms!) n as in (1) and $\underline{V}$-natural transformations $\eta: 1 \rightarrow TS$; n is the composite

$$\underline{A}(SB,A) \xrightarrow[T_{SB,A}]{} \underline{B}(TSB,TA) \xrightarrow[\underline{B}(\eta_B,1)]{} \underline{B}(B,TA), \qquad (2)$$

and η_B is the image of 1_{SB} under $Vn_{B,SB}$, where V denotes the basic functor $V: \underline{V} \longrightarrow Ens$. Similarly there is a bijection between $\underline{V}$-natural transformations $m: \underline{B}(B,TA) \longrightarrow \underline{A}(SB,A)$ and $\underline{V}$-natural transformations $\varepsilon: ST \rightarrow 1$. It is easy to verify that the composites mn, nm correspond respectively to $\varepsilon S \circ S\eta$ and $T\varepsilon \circ \eta T$, so that n, m are mutually inverse exactly when ε, η constitute a $\underline{V}$-adjunction $S \dashv T$.

We write $n; \varepsilon, \eta: S \dashv T: \underline{A}, \underline{B}$.

3.2. <u>In the situation of</u> 3.1, $S_{BB'}$ <u>is the composite</u>

$$\underline{B}(B,B') \xrightarrow[\underline{B}(1,\eta)]{} \underline{B}(B,TSB') \xrightarrow[n^{-1}]{} \underline{A}(SB,SB'). \qquad (3)$$

__Proof__. $S_{BB'}$ and (3) are both $\underline{V}$-natural, whence they coincide by the representation theorem provided that, when we apply V, put $B' = B$, and evaluate at 1_B, they give the same result. But $VS_{BB} \cdot 1 = 1$, while $Vn^{-1} \cdot V\underline{B}(1, \eta) \cdot 1 = Vn^{-1} \cdot \eta = 1$.

For a $\underline{V}$-category $\underline{B}$ and for $B \in \underline{B}$ we write $L^B : \underline{B} \to \underline{V}$ for the $\underline{V}$-functor represented by B, which sends B' to $\underline{B}(B, B')$.

3.3. __T: $\underline{A} \to \underline{B}$ has an adjoint if and only if, for each $B \in \underline{B}$, the $\underline{V}$-functor $L^B T : \underline{A} \to \underline{V}$ is representable.__

__Proof__. "Only if" is trivial. Given that $L^B T$ is representable, let SB be the representing object, and let (1) be the representation, so that (1) is an isomorphism $\underline{V}$-natural in A for each fixed B. Then (1) still determines η as in (2), and we define $S_{BB'}$ by (3). It is easily verified that S is a $\underline{V}$-functor and that n is $\underline{V}$-natural in B.

We leave the reader to verify:

3.4. __Suppose that we have in $\underline{V}$-$\underline{Cat}$ the situation of 2.1 above, with n as in (1) and n' correspondingly. Then the relation between λ and μ is the following: given λ, μ_B is that morphism (whose existence and uniqueness is guaranteed by the representation theorem) rendering commutative the diagram__

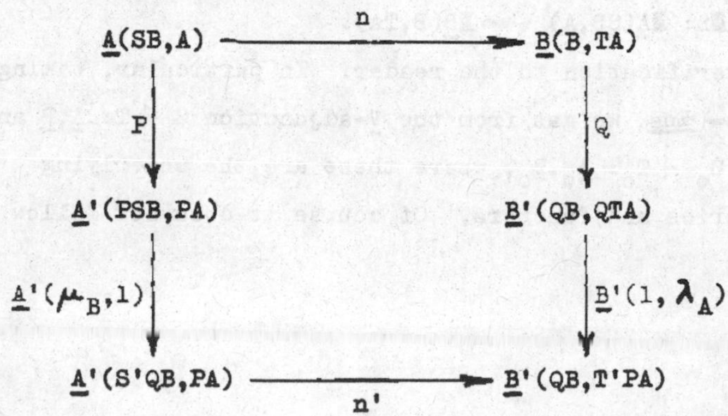

$$\begin{array}{ccc} \underline{A}(SB,A) & \xrightarrow{n} & \underline{B}(B,TA) \\ \downarrow P & & \downarrow Q \\ \underline{A}'(PSB,PA) & & \underline{B}'(QB,QTA) \\ \downarrow \underline{A}'(\mu_B,1) & & \downarrow \underline{B}'(1,\lambda_A) \\ \underline{A}'(S'QB,PA) & \xrightarrow{n'} & \underline{B}'(QB,T'PA) \end{array}$$

If we have $\underline{V}$-functors $T\colon \underline{C}^{op} \otimes \underline{A} \longrightarrow \underline{B}$ and $S\colon \underline{B} \otimes \underline{C} \longrightarrow \underline{A}$ with $n_C;\ \varepsilon_C, \eta_C\colon S(-C) \dashv T(C-)$ for each $C \in \underline{C}$, then each of n, ε, η is $\underline{V}$-natural in C if any one of them is; in which case we write $n;\ \varepsilon, \eta\colon S \dashv T$. In the extension of 3.4 to this case, λ is $\underline{V}$-natural in C if and only if μ is.

3.5. *Let* $T\colon \underline{C}^{op} \otimes \underline{A} \longrightarrow \underline{B}$ *and for each fixed* $C \in \underline{C}$ *let there be an adjunction* $n_C;\ \varepsilon_C, \eta_C\colon S(-C) \dashv T(C-)$. *Then there is a unique way of making* $S(-C)\colon \underline{B} \longrightarrow \underline{A}$ *into a* $\underline{V}$*-bifunctor* $S\colon \underline{B} \otimes \underline{C} \longrightarrow \underline{A}$ *so that we have* $n;\ \varepsilon, \eta\colon S \dashv T$.

Proof. If one writes the diagram expressing the $\underline{V}$-naturality in C of η, and uses (2), one sees that we are forced to define $S(B-)_{CC'}$ to be the composite

$$\begin{array}{ccc}
\underline{C}(CC') & & \underline{A}(S(BC),S(BC')) \\
{\scriptstyle T(-,S(BC'))}\downarrow & & \uparrow{\scriptstyle n^{-1}} \\
\underline{B}(T(C',S(BC')),\ T(C,S(BC'))) & \xrightarrow[\underline{B}(\eta,1)]{} & \underline{B}(B,T(C,S(BC')))
\end{array}$$

We leave to the reader the verification that S is a $\underline{V}$-bifunctor.

3.6. *Let* $\Phi\colon \underline{V} \longrightarrow \underline{V}'$ *be a closed functor, and let it induce the 2-functor* $\Phi_*\colon \underline{V}\text{-}\underline{Cat} \longrightarrow \underline{V}'\text{-}\underline{Cat}$. *Let* $n;\ \varepsilon, \eta\colon S \dashv T\colon \underline{A},\underline{B}$ *be a* $\underline{V}$*-adjunction and* $n';\ \Phi_*\varepsilon, \Phi_*\eta\colon \Phi_*S \dashv \Phi_*T\colon \Phi_*\underline{A}, \Phi_*\underline{B}$ *its image under* Φ_*. *Then* $n' = \Phi n\colon \Phi\underline{A}(SB,A) \longrightarrow \Phi\underline{B}(B,TA)$.

We leave this verification to the reader. In particular, taking Φ to be $V\colon \underline{V} \longrightarrow \underline{Ens}$, we get from the $\underline{V}$-adjunction $S \dashv T\colon \underline{A},\underline{B}$ an $\underline{Ens}$-adjunction $S_o \dashv T_o\colon \underline{A}_o,\underline{B}_o$, where these are the underlying ordinary categories and functors. Of course it does not follow that a $\underline{V}$-functor $T\colon \underline{A} \to \underline{B}$ has an adjoint merely because $T_o\colon \underline{A}_o \to \underline{B}_o$ has one.

4. Tensored and cotensored $\underline{V}$-categories. We say that the $\underline{V}$-category $\underline{A}$ is tensored if, for each $A \in \underline{A}$, the representable $\underline{V}$-functor $L^A\colon \underline{A} \to \underline{V}$ has a left adjoint. By 3.5 this means that $Hom_{\underline{A}}\colon \underline{A}^{op} \otimes \underline{A} \to \underline{V}$ has a left adjoint $Ten_{\underline{A}}\colon \underline{V} \otimes \underline{A} \to \underline{A}$. For simplicity we write $X \otimes A$ for $Ten_{\underline{A}}(X,A)$; we also abbreviate the internal-hom $\underline{V}(XY)$ of $\underline{V}$ to $[XY]$; so, for the present adjunction, (1) takes the form

$$p\colon \underline{A}(X \otimes A,B) \cong [X,\underline{A}(AB)]. \tag{4}$$

(Note that it is not enough for the underlying ordinary functor $\underline{A}(A-)\colon \underline{A}_o \to \underline{V}_o$ of L^A to have a left adjoint $- \otimes A\colon \underline{V}_o \to \underline{A}_o$, nor is it enough to have a natural isomorphism (4) which is not $\underline{V}$-natural.) Of course the $\underline{V}$-category $\underline{V}$ itself is tensored, since $p\colon [X \otimes A,B] \to [X[AB]]$ is $\underline{V}$-natural by ([3], page 543, Theorem 7.4).

When $\underline{A}$ is tensored we have natural isomorphisms $a\colon (X \otimes Y) \otimes A \to X \otimes (Y \otimes A)$, $\ell\colon I \otimes A \to A$, where $X, Y, I \in \underline{V}$ and $A \in \underline{A}$; these are most easily got by using 3.4 above. There are various "coherence" relations between the ℓ, a, p we have just introduced for $\underline{A}$ and the ℓ, a, p, r, i (where $r\colon X \otimes I \cong X$, $i\colon X \cong [IX]$) of $\underline{V}$ itself.

We say that the $\underline{V}$-category $\underline{A}$ is cotensored if the dual category $\underline{A}^{op}$ is tensored. We write $Cot_{\underline{A}}\colon \underline{V}^{op} \otimes \underline{A} \to \underline{A}$ for the $\underline{V}$-functor $(Ten_{\underline{A}^{op}})^{op}$, but denote $Cot_{\underline{A}}(X,A)$ by $[X,A]$ for short. We also write s for the p of $\underline{A}^{op}$, and so we have

$$s\colon \underline{A}(A,[XB]) \cong [X,\underline{A}(AB)]. \qquad (5)$$

The ℓ, a of $\underline{A}^{op}$ now become $B \cong [IB]$ and $[X \otimes Y,B] \cong [X[YB]]$, and there are more coherence relations. When $\underline{A}$ is both tensored and cotensored, there are still further coherence relations between its p and its s. The $\underline{V}$-category $\underline{V}$ itself is cotensored, with s being the composite

$$[Y,[XZ]] \xrightarrow{p^{-1}} [Y \otimes X,Z] \xrightarrow{[c,1]} [X \otimes Y,Z] \xrightarrow{p} [X,[YZ]].$$

In the case $\underline{V} = \underline{Ens}$, $X \otimes A$ is the coproduct and $[X,A]$ the product of X copies of A. For a general $\underline{V}$, then, we think of $X \otimes -$ as some kind of colimit and of $[X,-]$ as some kind of limit.

If $T\colon \underline{A} \to \underline{B}$ is a $\underline{V}$-functor, where $\underline{A}$ and $\underline{B}$ are cotensored $\underline{V}$-categories, there is by the representation theorem a unique morphism $\tau_{XA}\colon T[XA] \to [X,TA]$ rendering commutative the diagram

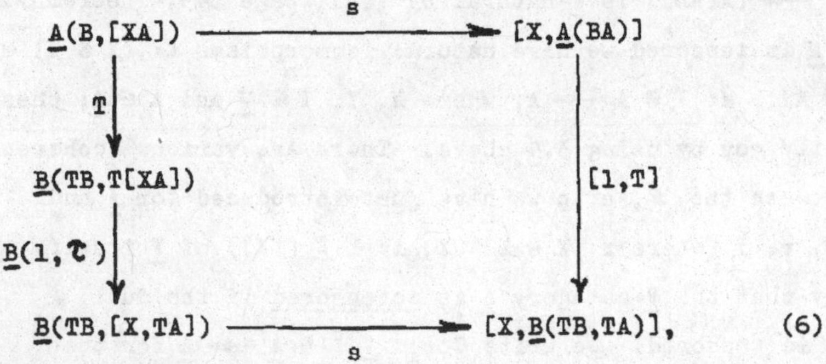

and τ is $\underline{V}$-natural because everything else is. If τ is an isomorphism we say that T <u>preserves cotensor products</u>.

4.1. <u>Let</u> $T\colon \underline{A} \to \underline{B}$ <u>be a</u> $\underline{V}$<u>-functor where</u> $\underline{A}$, $\underline{B}$ <u>are cotensored. In order for</u> T <u>to have a left adjoint it is necessary and sufficient that</u> (i) <u>the underlying ordinary functor</u> $T_o\colon \underline{A}_o \to \underline{B}_o$ <u>have a left adjoint, and</u> (ii) T <u>preserve cotensor products</u>.

<u>Proof</u>. Certainly it is necessary that T_o have an adjoint S_o, with $\varepsilon, \eta\colon S_o \dashv T_o$ say. Write SB for S_oB, and define n by (2); thus T has an adjoint if and only if n is an isomorphism. However from (2) and (6) we easily get the commutativity of

$$\begin{array}{ccccc}
\underline{B}_o(B,T[XA]) & \longrightarrow & \underline{A}_o(SB,[XA]) & \longrightarrow & \underline{V}_o(X,\underline{A}(SB,A)) \\
{\scriptstyle \underline{B}_o(1,\tau)}\downarrow & & & & \downarrow{\scriptstyle \underline{V}_o(1,n)} \\
\underline{B}_o(B,[X,TA]) & & \longrightarrow & & \underline{V}_o(X,\underline{B}(B,TA))
\end{array}$$

Since the horizontal arrows are isomorphisms, n is an isomorphism if and only if τ is.

There is a similar criterion for a $\underline{V}$-functor $T\colon \underline{A} \to \underline{V}$ to be representable. The condition that $\underline{B}$ be cotensored is not essential to 4.1, but then we must say differently what it means to preserve cotensor products.

5. <u>Tensored</u> $\underline{V}$<u>-categories and closed functors</u>. Recall from [3] that a closed functor $\Phi\colon \underline{V} \to \underline{V}'$ consists of a functor $\Phi\colon \underline{V}_o \to \underline{V}'_o$ together with natural transformations $\tilde{\Phi}\colon \Phi A \otimes \Phi B \to \Phi(A \otimes B)$ and $\Phi^o\colon I' \to \Phi I$ satisfying certain axioms. It induces a 2-functor $\Phi_*\colon \underline{V}\text{-}\underline{Cat} \to \underline{V}'\text{-}\underline{Cat}$. It also gives rise to a natural transformation $\hat{\Phi}\colon \Phi[AB] \to [\Phi A, \Phi B]$, which can be considered as giving a $\underline{V}'$-functor $\hat{\Phi}\colon \Phi_*\underline{V} \to \underline{V}'$ whose value $\hat{\Phi}A$ on objects is ΦA.

5.1. For a closed functor $\Phi: \underline{V} \to \underline{V}'$ the following assertions are equivalent:

(a) Whenever a $\underline{V}$-category $\underline{A}$ is tensored, so is the $\underline{V}'$-category $\Phi_*\underline{A}$.

(b) The $\underline{V}'$-category $\Phi_*\underline{V}$ is tensored.

(c) The $\underline{V}'$-functor $\hat{\Phi}: \Phi_*\underline{V} \to \underline{V}'$ has a left adjoint.

Proof. By ([3], page 449, (6.15)) $L'^A: \Phi_*\underline{A} \to \underline{V}'$ is the composite

$$\Phi_*\underline{A} \xrightarrow[\Phi_* L^A]{} \Phi_*\underline{V} \xrightarrow[\hat{\Phi}]{} \underline{V}'. \qquad (7)$$

Hence (c) implies (a) since if L^A has an adjoint so does $\Phi_* L^A$. That (a) implies (b) is trivial, and (b) implies (c) by taking $\underline{A} = \underline{V}$, $A = I$ in (7) and recalling that $L^I \cong 1$.

Recall that a closed functor $\Phi: \underline{V} \to \underline{V}'$ is called normal if $V'\Phi = V$, V being the basic closed functor $V: \underline{V} \to \underline{Ens}$ and V' similarly.

5.2. When $\Phi: \underline{V} \to \underline{V}'$ is a normal closed functor, the assertions (a), (b), (c) of 5.1 are also equivalent to:

(d) Φ has a left adjoint in the 2-category $\underline{C\ell}$ of closed categories, closed functors, and closed natural transformations.

Proof. Let $\varepsilon, \eta: \Psi \dashv \Phi: \underline{V}, \underline{V}'$ in $\underline{C\ell}$. We show that $\hat{\Phi}$ has a left adjoint whose value at A is ΨA. Define u as the composite

$$\Phi[\Psi A, B] \xrightarrow[\hat{\Phi}]{} [\Phi\Psi A, \Phi B] \xrightarrow[[\eta, 1]]{} [A, \Phi B]; \qquad (8)$$

certainly u is $\underline{V}'$-natural in B for each fixed A, and we have only to show that u is an isomorphism. Its inverse is v, defined as the composite

$$[A,\Phi B] \xrightarrow{\eta} \Phi\Psi[A,\Phi B] \xrightarrow{\Phi\hat{\Psi}} \Phi[\Psi A,\Psi\Phi B] \xrightarrow{\Phi[1,\varepsilon]} \Phi[\Psi A,B]. \quad (9)$$

The verification that u and v are mutually inverse is straightforward using (i) the definition of closed natural transformation on page 441 of [3], (ii) the naturality of η, ε, $\hat{\Phi}$, $\hat{\Psi}$, and (iii) the adjunction relations $\Phi\varepsilon\circ\eta\Phi = 1$, $\varepsilon\Psi\circ\Psi\eta = 1$. This part does not use the normality of Φ.

Now let $\hat{\Phi}$: $\Phi_*\underline{V} \to \underline{V}'$ have an adjoint S: $\underline{V}' \to \Phi_*\underline{V}$ with ε,η: S $\dashv \hat{\Phi}$. Because Φ is normal, $(\Phi_*\underline{V})_o = \underline{V}_o$ and $(\hat{\Phi})_o = \Phi\underline{V}_o \to \underline{V}'_o$. Define Ψ: $\underline{V}'_o \to \underline{V}_o$ to be S_o. The adjunction of S to $\hat{\Phi}$ corresponds to a natural isomorphism $(\Phi_*\underline{V})(\Psi A,B) \cong \underline{V}'[A,\Phi B]$, that is, $\Phi[\Psi A,B] \cong [A,\Phi B]$, which is $\underline{V}'$-natural in B. The composite isomorphism

$$(\Phi_*\underline{V})(\Psi(A \otimes B),C) \cong [A \otimes B,\Phi C] \cong [A,[B,\Phi C]] \cong [A,\Phi[\Psi B,C]]$$
$$\cong \Phi[\Psi A,[\Psi B,C]] \cong \Phi[\Psi A \otimes \Psi B,C] = (\Phi_*\underline{V})(\Psi A \otimes \Psi B,C)$$

corresponds by the representation theorem to a natural isomorphism $\tilde{\Psi}$: $\Psi A \otimes \Psi B \to \Psi(A \otimes B)$. Again, the composite isomorphism

$$(\Phi_*\underline{V})(\Psi I',A) \cong [I',\Phi A] \cong \Phi A \cong \Phi[IA] = (\Phi_*\underline{V})(I,A)$$

corresponds by the representation theorem to an isomorphism Ψ^o: I $\to \Psi$I'. It is easy to verify that Ψ, $\tilde{\Psi}$, Ψ^o constitute a closed functor Ψ: $\underline{V}' \to \underline{V}$, and that ε, η are in fact closed natural transformations, so that ε,η: $\Psi \dashv \Phi$ in $\underline{C\ell}$.

5.3. <u>If Φ satisfies the assertions of</u> 5.1, <u>and if the</u> $\underline{V}$-<u>functor</u> T: $\underline{A} \to \underline{B}$ <u>preserves tensor products, so does</u> Φ_*T: $\Phi_*\underline{A} \to \Phi_*\underline{B}$.

6. <u>The canonical decomposition of a closed functor</u>. Call a closed functor $\Phi: \underline{V} \to \underline{V}'$ <u>residual</u> if $\underline{V}$ and $\underline{V}'$ have the same objects, if Φ is the identity on objects, and if $\tilde{\Phi}: \Phi A \otimes \Phi B \to \Phi(A \otimes B)$, $\Phi^0: I' \to \Phi I$, $\hat{\Phi}: \Phi[AB] \to [\Phi A, \Phi B]$ are all 1.

6.1. <u>Every closed functor $\Phi: \underline{V} \to \underline{V}'$ has a unique factorization $\underline{V} \xrightarrow{\Phi_1} \overline{\underline{V}} \xrightarrow{\Phi_2} \underline{V}'$ where Φ_1 is residual and Φ_2 is normal.</u>

6.2. <u>Let $\alpha: \Phi \Rightarrow \Psi: \underline{V} \to \underline{V}'$ be a closed natural transformation, and let $\underline{V} \xrightarrow{\Phi_1} \overline{\underline{V}} \xrightarrow{\Phi_2} \underline{V}'$, $\underline{V} \xrightarrow{\Psi_1} \overline{\overline{\underline{V}}} \xrightarrow{\Psi_2} \underline{V}'$ be the factorizations of</u> Φ, Ψ. <u>Then there is uniquely a residual closed functor $\Pi: \overline{\underline{V}} \to \overline{\overline{\underline{V}}}$ and a closed natural transformation $\beta: \Phi_2 \to \Psi_2\Pi$ such that $\Pi\Phi_1 = \Psi_1$ and $\beta\Phi_1 = \alpha$.</u>

We omit the proofs of 6.1, 6.2. If Φ has the factorization in 6.1, the $\underline{V}'$-functor $\hat{\Phi}: \Phi_*\underline{V} \to \underline{V}'$ is identical with $\hat{\Phi}_2: \Phi_{2*}\overline{\underline{V}} \to \underline{V}'$. Hence we can replace 5.2 by:

6.3. <u>For a closed functor $\Phi: \underline{V} \to \underline{V}'$, the assertions of</u> (5.1) <u>are equivalent to</u>

(e) <u>If Φ has the factorization $\Phi_2\Phi_1$, Φ_2 has a left adjoint in the</u> 2-<u>category</u> $\mathcal{C}\ell$.

The reader should note that the adjoint Ψ of Φ in (5.2) is in general neither normal nor residual.

References

[1] J. Beck, On H-spaces and infinite loop spaces. (Preprint)

[2] Marta Bunge, Relative functor categories and categories of algebras. Journal of Algebra 11 (1969) 64-101.

[3] S. Eilenberg and G. M. Kelly, Closed categories. Proc. Conf. Categorical Algebra (La Jolla 1965). Springer-Verlag 1966.

[4] A. Kock, Closed categories generated by commutative monads. Aarhus Universitet Matematisk Institut Preprint Series 1968/69 No. 13.

The University of New South Wales

ENRICHED FUNCTOR CATEGORIES

by B. J. Day and G. M. Kelly

Received April 30, 1969

1. *Introduction*. For a closed category $\underline{V}$ and $\underline{V}$-categories $\underline{A}$, $\underline{B}$ one would like to exhibit a $\underline{V}$-category $[\underline{A},\underline{B}]$ whose underlying ordinary category would be the category of $\underline{V}$-functors from $\underline{A}$ to $\underline{B}$ and $\underline{V}$-natural transformations between them. We present here some observations on the problem. The remarks in the introduction of the preceding paper [3] apply also here. In the case $\underline{B} = \underline{V}$ the $\underline{V}$-category $[\underline{A},\underline{V}]$ is constructed by Marta Bunge in her thesis [1]. Much that we do below is well known in the cases $\underline{V} = \underline{Ens}$ or $\underline{V} = \underline{Ab}$; see in particular Ulmer [4]. What we call *ends* and *coends* were introduced in the case $\underline{V} = \underline{Ab}$ by Yoneda [5]; we borrow from him the "integral" notation. Some of the matters below have been discussed in correspondence between us and Ross Street. Finally, we are informed by Mac Lane that Bénabou's unpublished results are very similar to ours.

We assume familiarity with the preceding paper [3]. As there, we use the term *closed category* to mean what was called in [2] a *symmetric monoidal closed category*.

2. *Limits in $\underline{V}$-categories*. Let $\underline{V}$ be a closed category, $\underline{A}$ a $\underline{V}$-category, and $F\colon \underline{K} \to \underline{A}_o$ an ordinary functor from a category $\underline{K}$ to the underlying category $\underline{A}_o$ of $\underline{A}$. A family $\alpha_K\colon M \to FK$ $(K \in \underline{K})$ of morphisms in $\underline{A}_o$, which is the limit of F in the usual sense,

will be called <u>the limit of</u> F <u>in</u> $\underline{A}_o$. The family α_K is called <u>the limit of</u> F <u>in</u> $\underline{A}$ if, for each $A \in \underline{A}$, the family $\underline{A}(A,\alpha_K)$: $\underline{A}(A,M) \longrightarrow \underline{A}(A,FK)$ is a limit in $\underline{V}_o$. In this case, applying the representable basic functor V: $\underline{V}_o \longrightarrow \underline{Ens}$, we see that $\underline{A}_o(A,M) \longrightarrow \underline{A}_o(A,FK)$ is a limit in $\underline{Ens}$, so that $M \longrightarrow FK$ is certainly a limit in $\underline{A}_o$. However limits in $\underline{A}_o$ need not be limits in $\underline{A}$. They will be so if, for each $A \in \underline{A}$, the functor $\underline{A}(A-)$: $\underline{A}_o \longrightarrow \underline{V}_o$ preserves limits. This is automatically the case if $\underline{V} = \underline{Ens}$ or $\underline{Ab}$; it is also the case wherever $\underline{A}$ is tensored (for then the representable $\underline{V}$-functor L^A: $\underline{A} \longrightarrow \underline{V}$ has an adjoint, so that its underlying functor $\underline{A}(A-)$: $\underline{A}_o \longrightarrow \underline{V}_o$ has one too).

We describe a $\underline{V}$-category $\underline{A}$ as <u>complete</u> if

(i) the underlying ordinary category $\underline{A}_o$ is complete;

(ii) for each $A \in \underline{A}$ the functor $\underline{A}(A-)$: $\underline{A}_o \longrightarrow \underline{V}_o$ preserves small limits;

(iii) $\underline{A}$ is cotensored.

We define <u>cocompleteness</u> dually. Since $\underline{V}$ itself is tensored and cotensored, it is complete precisely when $\underline{V}_o$ is.

It is easy to verify:

2.1. <u>If $\underline{A}$ has cotensor products then L^A: $\underline{A} \longrightarrow \underline{V}$ preserves them for each $A \in \underline{A}$.</u>

3. <u>Ends and coends</u>. Let $\underline{V}$ be a closed category, let $\underline{A}$ and $\underline{B}$ be $\underline{V}$-categories, and let T: $\underline{A}^{op} \otimes \underline{A} \longrightarrow \underline{B}$ be a $\underline{V}$-functor. By an <u>end</u> of T is meant an object K of $\underline{B}$ and a family γ_A: $K \longrightarrow T(AA)$ of morphisms in $\underline{B}_o$ such that (i) γ is $\underline{V}$-natural, and (ii) any

$\underline{V}$-natural family $\delta_A : N \to T(AA)$ is of the form $\delta_A = \gamma_A \lambda$ for a unique $\lambda : N \to K$. Clearly an end is unique (to within a unique isomorphism) when it exists. Similarly we define a coend of T to be a $\underline{V}$-natural family $\rho_A : T(AA) \to Q$ which is initial among all such.

3.1. The end of $T: \underline{A}^{op} \otimes \underline{A} \to \underline{B}$ exists if $\underline{A}$ is small, the underlying category $\underline{B}_o$ of $\underline{B}$ is complete, and $\underline{B}$ is cotensored.

Proof. The criterion for $\underline{V}$-naturality of $\delta_A : N \to T(AA)$ is the commutativity of

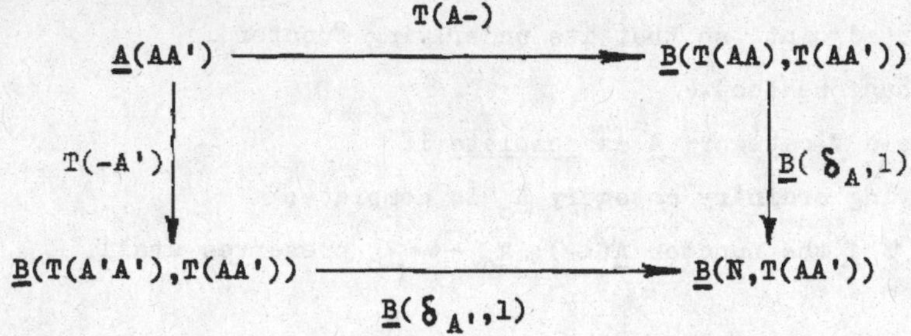

Under the cotensoring isomorphism $\sigma : \underline{V}_o(X, \underline{B}(PQ)) \cong \underline{B}_o(P, [XQ])$ this becomes

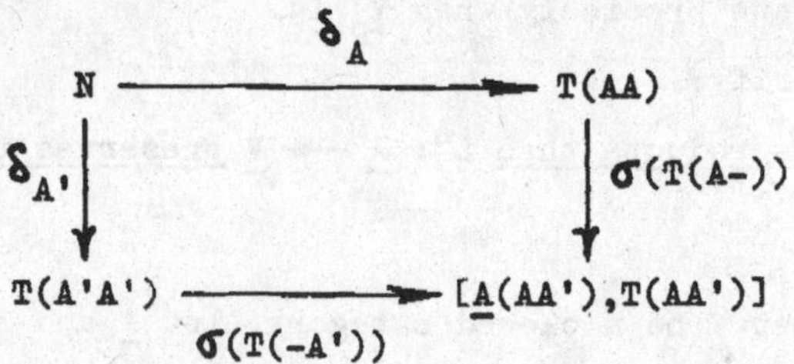

It is clear that the terminal such δ is precisely the limit of a diagram in $\underline{B}_o$ which is small if $\underline{A}$ is small; hence this exists by the completeness of $\underline{B}_o$.

The same analysis allows us to prove:

3.2. Let $\underline{A}$ be a $\underline{V}$-category (resp. small $\underline{V}$-category), let $T: \underline{A}^{op} \otimes \underline{A} \to \underline{B}$ be a $\underline{V}$-functor and let $\gamma_A: K \to T(AA)$ be its end. Suppose that $\underline{B}$ is cotensored. Let the $\underline{V}$-functor $P: \underline{B} \to \underline{C}$ preserve cotensor products and let its underlying functor $P_o: \underline{B}_o \to \underline{C}_o$ preserve limits (resp. small limits). Then $P\gamma_A: PK \to PT(AA)$ is the end of PT. In particular this is the case if P has a left adjoint.

What we have called above the end of T will now be called more precisely the end of T in $\underline{B}_o$. We shall say that $\gamma_A: K \to T(AA)$ is the end of T in $\underline{B}$ if, for each $B \in \underline{B}$, $\underline{B}(B, \gamma_K): \underline{B}(BK) \to \underline{B}(B, T(AA))$ is the end in $\underline{V}_o$ of $L^B T$. Then γ is certainly the end of T in $\underline{B}_o$, for there is a bijection between $\underline{V}$-natural families $I \to \underline{B}(B, T(AA))$ and $\underline{V}$-natural families $B \to T(AA)$. Combining 3.1 and 3.2 and using 2.1 gives:

3.3. The end in $\underline{B}$ of $T: \underline{A}^{op} \otimes \underline{A} \to \underline{B}$ exists if $\underline{A}$ is small and $\underline{B}$ is complete.

From 3.2 we also get:

3.4. The end in $\underline{B}_o$ of $T: \underline{A}^{op} \otimes \underline{A} \to \underline{B}$ is an end in $\underline{B}$ if $\underline{B}$ is tensored.

If $\gamma_A: K \to T(AA)$ is the end in $\underline{B}$ of T we write $\int_A T(AA)$ for K, and we call γ_A the A'th projection. This notation will not be used if γ is only the end in $\underline{B}_o$ and not in $\underline{B}$. Similarly we write $\int^A T(AA)$ for the coend in $\underline{B}$. Since $\underline{V}$ is tensored we can by 3.4 always write

$$\underline{B}(B, \int_A T(AA)) \cong \int_A \underline{B}(B, T(AA)).$$

For a $\underline{V}$-functor P: $\underline{B} \to \underline{C}$, if T: $\underline{A}^{op} \otimes \underline{A} \to \underline{B}$ has an end $\int_A T(AA)$, then γ_A: $\int_A T(AA) \to T(AA)$ is $\underline{V}$-natural whence $P\gamma_A$: $P\int_A T(AA) \to PT(AA)$ is also $\underline{V}$-natural. Thus we get a canonical morphism $P\int_A T(AA) \to \int_A PT(AA)$ if both sides exist.

If α: T $\to$ S: $\underline{A}^{op} \otimes \underline{A} \to \underline{B}$ then $\int_A T(AA) \xrightarrow[\gamma_A]{} T(AA) \xrightarrow[\alpha_{AA}]{} S(AA)$ is $\underline{V}$-natural in A and induces a morphism $\int\alpha$: $\int_A T(AA) \to \int_A S(AA)$ if these ends exist.

If T: $\underline{A}^{op} \otimes \underline{A} \otimes \underline{C} \to \underline{B}$ and if $\int_A T(AAC)$ exists for each C, then the composite

$$\underline{C}(CC') \xrightarrow[T(AA-)]{} \underline{B}(T(AAC),T(AAC')) \xrightarrow[\underline{B}(\gamma_A,1)]{} \underline{B}(\int_A T(AAC),T(AAC')),$$

being $\underline{V}$-natural in A, induces morphisms

$$\underline{C}(CC') \longrightarrow \int_A \underline{B}(\int_A T(AAC),T(AAC')) \cong \underline{B}(\int_A T(AAC),\int_A T(AAC')),$$

which give to $\int_A T(AAC)$ the structure of a $\underline{V}$-functor $\underline{C} \to \underline{B}$.

If T: $\underline{A}^{op} \otimes \underline{B}^{op} \otimes \underline{A} \otimes \underline{B} \to \underline{C}$, suppose that $\int_B T(ABA'B)$ exists for each A, A'; this is then a $\underline{V}$-functor $\underline{A}^{op} \otimes \underline{A} \to \underline{C}$; suppose further that $\int_A \int_B T(ABAB)$ exists. Then it follows at once that this is also the end $\int_{A,B} T(ABAB)$ of T considered as a $\underline{V}$-functor $(\underline{A} \otimes \underline{B})^{op} \otimes (\underline{A} \otimes \underline{B}) \to \underline{C}$. We conclude that $\int_A \int_B T(ABAB) \cong \int_B \int_A T(ABAB)$ wherever all these ends exist in $\underline{C}$.

3.5. <u>Let T: $\underline{A} \to \underline{B}$ be a $\underline{V}$-functor and let $\underline{B}$ be tensored. Then $\int^A \underline{A}(AB) \otimes TA = TB$, the coprojection τ: $\underline{A}(AB) \otimes TA \to TB$ corresponding by adjunction to T_{AB}: $\underline{A}(AB) \to \underline{B}(TA,TB)$.</u>

<u>Proof</u>. We have to show that, for each $C \in \underline{B}$, $\underline{B}(\tau,C)$: $\underline{B}(TB,C) \to \underline{B}(\underline{A}(AB) \otimes TA,C)$ is an end in $\underline{V}_o$. But

$\underline{B}(\underline{A}(AB) \otimes TA, C) \cong [\underline{A}(AB), \underline{B}(TA,C)]$ by a $\underline{V}$-natural isomorphism. Let $\delta_A : X \to [\underline{A}(AB), \underline{B}(TA,C)]$ be $\underline{V}$-natural in A; then so is the corresponding $\varepsilon_A : \underline{A}(AB) \to [X, \underline{B}(TA,C)]$. By the representation theorem, then, ε_A is of the form

$$\underline{A}(AB) \xrightarrow[T_{AB}]{} \underline{B}(TA,TB) \xrightarrow[R^C]{} [\underline{B}(TB,C), \underline{B}(TA,C)] \xrightarrow[[\theta,1]]{} [X, \underline{B}(TA,C)]$$

for a unique $\theta : X \to \underline{B}(TB,C)$. This is the transform under adjunction of the assertion that δ_A is of the form

$$X \xrightarrow[\theta]{} \underline{B}(TB,C) \xrightarrow[\underline{B}(\tau,C)]{} \underline{B}(\underline{A}(AB) \otimes TA, C) \cong [\underline{A}(AB), \underline{B}(TA,C)]$$

for a unique θ; which is what we want.

4. <u>Functor categories</u>. Henceforth we take $\underline{V}$ to be a closed category with $\underline{V}_o$ complete: so that as a $\underline{V}$-category $\underline{V}$ is complete.

4.1. *If $\underline{A}$, $\underline{B}$ are $\underline{V}$-categories with $\underline{A}$ small, we can find, uniquely to within isomorphism, a $\underline{V}$-category $\underline{F}$ and $\underline{V}$-functors $E^A : \underline{F} \to \underline{B}$ $(A \in \underline{A})$, such that*

(i) *the objects of $\underline{F}$ are the $\underline{V}$-functors $\underline{A} \to \underline{B}$;*

(ii) *the value of E^A at $T \in \underline{F}$ is $TA \in \underline{B}$;*

(iii) *the family $E^A_{TS} : \underline{F}(TS) \to \underline{B}(TA,SA)$ is the end of the $\underline{V}$-functor*

$$\underline{A}^{op} \otimes \underline{A} \xrightarrow[T^{op} \otimes S]{} \underline{B}^{op} \otimes \underline{B} \xrightarrow[\mathrm{Hom}]{} \underline{V}.$$

<u>Proof</u>. (i) and (ii) tell us what the objects of $\underline{F}$ and the value on objects of E^A are to be; we define $\underline{F}(TS)$ as $\int_A \underline{B}(TA,SA)$ with E^A_{TS} as the projections, this being possible because $\underline{A}$ is small and

$\underline{V}$ is complete. If E^A is to be a $\underline{V}$-functor the other elements of structure j, M of $\underline{F}$ must be such as to render commutative

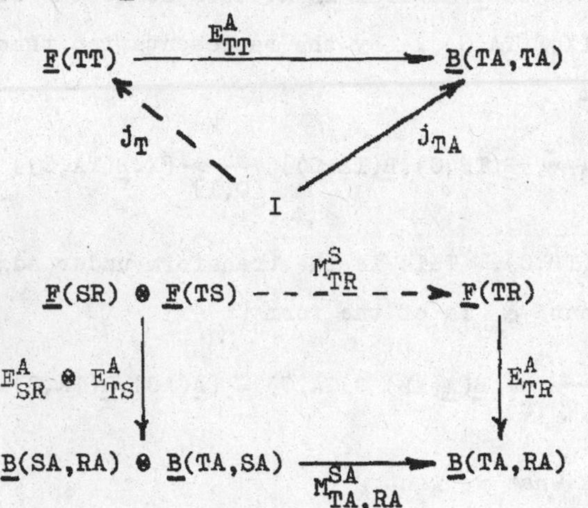

These diagrams serve to define j_T, M^S_{TR} uniquely precisely because E^A is an end. There remains only the easy verification that $\underline{F}$ satisfies the axioms for a $\underline{V}$-category.

We write $[\underline{A},\underline{B}]$ for $\underline{F}$, and call it a __functor__ __category__. The underlying ordinary category $[\underline{A},\underline{B}]_0$ has the same objects, while a morphism $T \longrightarrow S$ in $[\underline{A},\underline{B}]_0$ is essentially a morphism $I \longrightarrow \underline{F}(TS)$ in $\underline{V}_0$. These are in bijection with $\underline{V}$-natural families $I \longrightarrow \underline{B}(TA,SA)$, since $\underline{F}(TS)$ is an end; and these families are in bijection with $\underline{V}$-natural families $TA \longrightarrow SA$; so that $[\underline{A},\underline{B}]_0$ is the ordinary category of $\underline{V}$-functors and $\underline{V}$-natural transformations.

Define a $\underline{V}$-functor $H: [\underline{A},\underline{B}] \otimes \underline{A} \longrightarrow \underline{B}$ by $H(T-) = T: \underline{A} \longrightarrow \underline{B}$ and $H(-A) = E^A: [\underline{A},\underline{B}] \longrightarrow \underline{B}$. This will be a $\underline{V}$-functor because T and E^A are and because E^A is $\underline{V}$-natural in A([2], page 541 Proposition 7.1). It is easy to see that the diagram

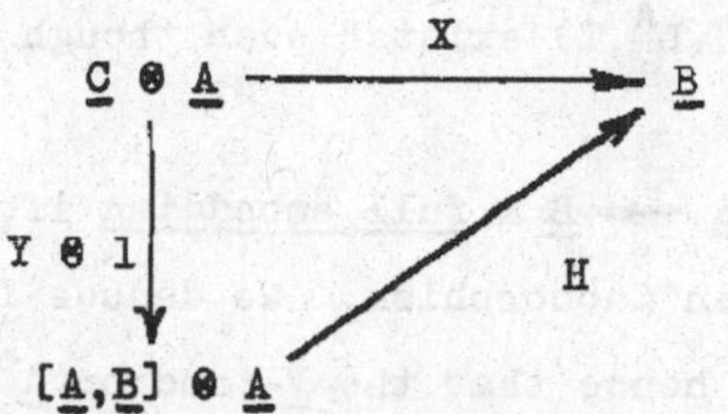

sets up a bijection between V-functors Y: C ⟶ [A,B] and V-functors X: C ⊗ A ⟶ B. Thus the operations A ⊗ B and [A,B] turn the category of small V-categories into a <u>closed</u> category <u>V-Cat</u>. In fact the structure of <u>V-Cat</u> is richer than this: it is a closed <u>2-category</u>, and has also a <u>duality structure</u> given by the involution $A \leftrightarrow A^{op}$. The detailed description of these structures would be out of place in the present summary.

5. <u>The higher representation theorem</u>. The representation theorem for V-categories proved in [2] establishes a bijection between V-natural transformations $\alpha: L^A \to T: A \to V$ and elements of the set VTA, where $V: V_o \to$ <u>Ens</u> is the basic functor. We can now do better:

5.1. <u>For a small V-category A there is a V-natural</u> (<u>in</u> T <u>and</u> A) <u>isomorphism</u>

$$A = A_{T,A}: TA \to [A,V](L^A,T).$$

<u>Proof</u>. $[A,V](L^A,T) = \int_B [L^A B, TB] = \int_B [A(AB),TB] = TA$ by 3.5.

We leave the reader to verify that A is in fact V-natural when seen as an isomorphism between two V-functors $[A,V] \otimes A \to V$.

There is a sense in which 5.1 is true whether A is small

or not; one can say that "$[\underline{A},\underline{V}](L^A,T)$ exists" even though $[\underline{A},\underline{V}](S,T)$ may not exist for a general S.

We call a $\underline{V}$-functor $T: \underline{A} \longrightarrow \underline{B}$ a <u>full embedding</u> if each $T_{AA'}: \underline{A}(AA') \longrightarrow \underline{B}(TA,TA')$ is an isomorphism. We deduce from 5.1 that $[\underline{A},\underline{V}](L^A,L^B) \cong \underline{A}(BA)$, and hence that the $\underline{V}$-functor $\underline{A}^{op} \longrightarrow [\underline{A},\underline{V}]$, sending A to L^A and corresponding by adjunction to the $\underline{V}$-functor Hom: $\underline{A}^{op} \otimes \underline{A} \longrightarrow \underline{V}$, is a full embedding; we call it the <u>Yoneda embedding</u>.

Given a $\underline{V}$-functor $P: \underline{A} \longrightarrow \underline{V}$ and an object B of the tensored $\underline{V}$-category $\underline{B}$, we have a $\underline{V}$-functor $\underline{A} \longrightarrow \underline{B}$ sending $A \in \underline{A}$ to $PA \otimes B$; we call this $\underline{V}$-functor $P \otimes B$. If we also have $T: \underline{A} \longrightarrow \underline{B}$, it is clear that $[\underline{A},\underline{B}](P \otimes B,T) \cong [\underline{A},\underline{V}](P,L^BT)$, since $\int_A \underline{B}(PA \otimes B,TA) = \int_A [PA,\underline{B}(B,TA)]$. If we take P to be $L^A: \underline{A} \longrightarrow \underline{V}$, we find that $[\underline{A},\underline{B}](L^A \otimes B,T) \cong [\underline{A},\underline{V}](L^A,L^BT) \cong \underline{B}(B,TA)$ by 5.1. Ulmer calls $L^A \otimes B$ a <u>generalized representable functor</u>, and the above result then appears as a generalized higher representation theorem.

It is easy to see that the functor category $[\underline{A},\underline{B}]$ inherits whatever good properties $\underline{B}$ has; in particular it is complete if $\underline{B}$ is, and limits, cotensor products, and ends in $[\underline{A},\underline{B}]$ are formed termwise. We can therefore write 3.5 for a tensored $\underline{B}$ as

5.2. $\int^A L^A \otimes TA = T,$

and interpret this as a canonical expression of a general $T: \underline{A} \longrightarrow \underline{B}$ as a coend of generalized representable functors.

6. <u>Kan Adjoints</u>

6.1. <u>Let</u> $P: \underline{C} \longrightarrow \underline{A}$ <u>be a</u> $\underline{V}$<u>-functor where</u> $\underline{A}$, $\underline{C}$ <u>are small, and let</u>

$\underline{B}$ be a cocomplete $\underline{V}$-category. Then the $\underline{V}$-functor $[P,1]: [\underline{A},\underline{B}] \to [\underline{C},\underline{B}]$ *has the left adjoint* Q, *where for* $S \in [\underline{C},\underline{B}]$ *the $\underline{V}$-functor* $Q(S): \underline{A} \to \underline{B}$ *is*

$$Q(S) = \int^C L^{PC} \otimes SC.$$

Proof. For any $T \in [\underline{A},\underline{B}]$ we have

$$[\underline{A},\underline{B}](Q(S),T) = \int_A \underline{B}(Q(S)A,TA) = \int_A \underline{B}(\int^C \underline{A}(PC,A) \otimes SC,TA)$$

$$= \int_A \int_C \underline{B}(\underline{A}(PC,A) \otimes SC,TA) = \int_A \int_C [\underline{A}(PC,A),\underline{B}(SC,TA)]$$

$$= \int_C \underline{B}(SC,TPC) \text{ by 3.5 } = [\underline{C},\underline{B}](S,TP) = [\underline{C},\underline{B}](S,[P,1]T).$$

We call Q the *Kan adjoint* of $[P,1]$.

6.2. *If, in* 6.1, $\underline{C}$ *is a full subcategory of* $\underline{A}$ *and* P *is the inclusion, then* $Q(S): \underline{A} \to \underline{B}$ *is an extension of* $S: \underline{C} \to \underline{B}$.

Proof. For $C \in \underline{C}$, the composite of $P: \underline{C} \to \underline{A}$ and $L^{PC}: \underline{A} \to \underline{V}$ is $L^C: \underline{C} \to \underline{V}$, because $\underline{C}$ is a full subcategory. So

$$Q(S)P = \int^C L^{PC}P \otimes SC = \int^C L^C \otimes SC = S \text{ by 5.2.}$$

We call $Q(S)$ in this case the *Kan extension* of S.

We get an important special case of 6.2 by taking $\underline{A}$ as the functor category $[\underline{C}^{op},\underline{V}]$ and P as the Yoneda embedding $\underline{C} \to [\underline{C}^{op},\underline{V}]$ sending $C \in \underline{C}$ to the right represented functor $R^C: \underline{C}^{op} \to \underline{V}$. In this case $L^{PC} \in [\underline{C}^{op},\underline{V}]$ is isomorphic by 5.1 to the evaluation functor $E^C: \underline{C}^{op} \to \underline{V}$, and therefore the Kan extension $Q(S): [\underline{C}^{op},\underline{V}] \to \underline{B}$ of $S: \underline{C} \to \underline{B}$ is given by $Q(S) = \int^C E^C \otimes SC$. Its value at $T \in [\underline{C}^{op},\underline{V}]$ is therefore $Q(S)(T) = \int^C TC \otimes SC$.

6.3. <u>The above Kan extension</u> Q(S): $[\underline{C}^{op},\underline{V}] \longrightarrow \underline{B}$ <u>has a right adjoint whose value at</u> $B \in \underline{B}$ <u>is the</u> $\underline{V}$-<u>functor</u> $\underline{C}^{op} \xrightarrow[S^{op}]{} \underline{B}^{op} \xrightarrow[R^B]{} \underline{V}$. <u>Therefore</u> Q(S) <u>preserves colimits and preserves tensor products</u>. <u>If</u> Z: $[\underline{C}^{op},\underline{V}] \longrightarrow \underline{B}$ <u>is any other extension of</u> S <u>that preserves small colimits and preserves tensor products</u>, Z <u>is isomorphic to</u> Q(S).

<u>Proof</u>. To see that Q(S) has the given right adjoint, let $T \in [\underline{C}^{op},\underline{V}]$. Then

$$\underline{B}(Q(S)T,B) = \underline{B}(\textstyle\int^C TC \otimes SC,B) = \int_C \underline{B}(TC \otimes SC,B)$$
$$= \int_C [TC,\underline{B}(SC,B)] = [\underline{C}^{op},\underline{V}](T,R^B S^{op}).$$

Let Z have the desired properties. Then, since $T = \int^C R^C \otimes TC$ by 5.2, we must have $Z(T) = \int^C Z(R^C \otimes TC)$ by 3.2. However it is easy to see that, in $[\underline{C}^{op},\underline{V}]$, $R^C \otimes TC$ is just the tensor product $TC \otimes R^C$; hence $Z(T) = \int^C TC \otimes Z(R^C)$. To say that Z extends S is to say that $Z(R^C) = SC$, so that Z coincides with Q(S).

7. <u>Adequacy</u>. Let S: $\underline{C} \longrightarrow \underline{B}$ be the inclusion in the $\underline{V}$-category $\underline{B}$ of a full subcategory $\underline{C}$. Following Isbell's terminology, we call $\underline{C}$ <u>adequate</u> in $\underline{B}$ if for each A, $B \in \underline{B}$ we have $\int_C[\underline{B}(SC,A),\underline{B}(SC,B)] \cong \underline{B}(AB)$. It follows from 3.5 that $\underline{B}$ is adequate in itself. If the full subcategory $\underline{C}$ of $\underline{B}$ is small, we have a $\underline{V}$-functor N: $\underline{B} \longrightarrow [\underline{C}^{op},\underline{V}]$ sending $B \in \underline{B}$ to the $\underline{V}$-functor $\underline{C}^{op} \xrightarrow[S^{op}]{} \underline{B}^{op} \xrightarrow[R^B]{} \underline{V}$; to say that $\underline{C}$ is adequate is precisely to say that N is a full embedding.

7.1. <u>Let</u> $\underline{B}$ <u>be cocomplete and</u> $\underline{C}$ <u>small. Then</u> $\underline{C}$ <u>is adequate in</u> $\underline{B}$ <u>if and only if every</u> $\underline{V}$-<u>natural family</u> α_C: $\underline{B}(SC,A) \longrightarrow \underline{B}(SC,B)$ <u>is of the form</u> $\underline{B}(1,f)$ <u>for a unique</u> f: $A \longrightarrow B$ <u>in</u> $\underline{B}_o$.

Proof. The condition for adequacy may be written, since $\underline{B}$ is tensored, in the form $\int_C \underline{B}(\underline{B}(SC,A) \otimes SC,B) \cong \underline{B}(AB)$. Since $\underline{B}$ is cocomplete this is equivalent by 3.2 to the assertion that A is the coend in $\underline{B}_o$ of $\underline{B}(S-,A) \otimes S-$. Since there is a bijection between $\underline{V}$-natural families $\underline{B}(SC,A) \otimes SC \to B$ and $\underline{V}$-natural families $\underline{B}(SC,A) \to \underline{B}(SC,B)$, this reduces to the condition in the theorem.

If $\underline{C}$ is small and $\underline{B}$ is cocomplete, the functor N: $\underline{B} \to [\underline{C}^{op},\underline{V}]$ is by 6.3 the right adjoint of the Kan extension Q(S): $[\underline{C}^{op},\underline{V}] \to \underline{B}$ of S: $\underline{C} \to \underline{B}$. So if $\underline{C}$ is adequate in $\underline{B}$, $\underline{B}$ is a full reflexive subcategory of $[\underline{C}^{op},\underline{V}]$. Since it is easy to see that a full reflexive subcategory of a cocomplete $\underline{V}$-category is itself cocomplete, and since $\underline{C}$ is clearly adequate in $[\underline{C}^{op},\underline{V}]$, we have:

7.2. Let $\underline{V}_o$ be cocomplete as well as complete. Then $\underline{B}$ is a full reflexive subcategory of $[\underline{C}^{op},\underline{V}]$ for some small $\underline{C}$ if and only if $\underline{B}$ is cocomplete and $\underline{B}$ has a small adequate subcategory.

We leave it to the reader to prove:

7.3. Let $\underline{V}_o$ be cocomplete as well as complete. Then $\underline{B}$ is equivalent to $[\underline{C}^{op},\underline{V}]$ for some small $\underline{C}$ if and only if $\underline{B}$ is cocomplete and has a small adequate subcategory $\underline{D}$ such that, for each $D \in \underline{D}$, L^D: $\underline{B} \to \underline{V}$ preserves small colimits and tensor products.

8. Change of closed category. Let Φ: $\underline{V} \to \underline{V}'$ be a closed functor, where $\underline{V}_o, \underline{V}'_o$ are complete. It was shown in [2] that Φ induces a monoidal functor (in fact 2-functor) $\Phi_\#$: $\underline{V}_\# \to \underline{V}'_\#$, where $\underline{V}_\# = \underline{V}$-$\underline{Cat}$. Now that we know $\underline{V}_\#$ to be closed, it follows that $\Phi_\#$ is a closed functor. Similarly a closed natural transformation

$\varepsilon\colon \Phi \longrightarrow \Psi$ gives a closed natural transformation $\varepsilon_{\#}\colon \Phi_{\#} \longrightarrow \Psi_{\#}$, and $(\ )_{\#}$ is a 2-functor.

Let Φ be normal, and let the $\underline{V}'$-functor $\hat{\Phi}\colon \Phi_{*}\underline{V} \longrightarrow \underline{V}'$ have a left adjoint. Then we saw in [3] that Φ has a left adjoint $\varepsilon, \eta\colon \Psi \dashv \Phi$ in the 2-category $\underline{\mathcal{C}\ell}$ of closed categories. It follows that $\varepsilon_{\#}, \eta_{\#}\colon \Psi_{\#} \dashv \Phi_{\#}$ in $\underline{\mathcal{C}\ell}$, so that by [3] the $\underline{V}'_{\#}$-category $(\Phi_{\#})_{*}\underline{V}_{\#}$ is tensored and cotensored. So for a $\underline{V}'$-category $\underline{X}$ and a $\underline{V}$-category $\underline{A}$ we have $\underline{V}$-categories $\underline{X} \otimes' \underline{A}$ and $[\underline{X}, \underline{A}]'$. In terms of the $\otimes$ and $[\ ,\ ]$ of $\underline{V}_{\#}$ these are easily seen to be $\Psi_{*}\underline{X} \otimes \underline{A}$ and $[\Psi_{*}\underline{X}, \underline{A}]$. The objects of this latter are the $\underline{V}$-functors $\Psi_{*}\underline{X} \longrightarrow \underline{A}$, but these are in bijection with the $\underline{V}'$-functors $\underline{X} \longrightarrow \Phi_{*}\underline{A}$. Thus we have succeeded in making these last into a $\underline{V}$-category.

9. Closed functor categories. Our functor categories have so far been $\underline{V}$-categories but never themselves closed categories. If $\underline{V}$ is a closed category and $\underline{A}$ a small $\underline{V}$-category, and if we want $[\underline{A}, \underline{V}]$ to be a closed category, we need to define for each $T \in [\underline{A}, \underline{V}]$ a functor $- \otimes T\colon [\underline{A}, \underline{V}] \longrightarrow [\underline{A}, \underline{V}]$ which has a right adjoint. We can by 6.3 get $- \otimes T$ as a Kan extension if we know what $L^{A} \otimes T$ is to be. Since we want $\otimes$ to be symmetric, it suffices to know $L^{A} \otimes L^{B}$. So for each $A, B \in \underline{A}$ we want a functor $\underline{A} \longrightarrow \underline{V}$; this requires a certain structure on $\underline{A}$. The examination of the necessary structure and the construction of closed functor categories will be the subject of a forthcoming paper by B. J. Day.

References

[1] Bunge, Marta C., Relative functor categories and categories of algebras. To appear in Journal of Algebra 11 (1969) 64-101.

[2] Eilenberg, S. and Kelly, G. M., Closed categories. Proc. Conf. on Categorical Algebra (La Jolla 1965), (Springer-Verlag 1966), 421 - 562.

[3] Kelly, G. M., Adjunction for enriched categories. These reports, supra.

[4] Ulmer, F., Representable functors with values in arbitrary categories. Journal of Algebra 8 (1968), 96 - 129.

[5] Yoneda, N., On Ext and exact sequences. Jour. Fac. Sci. Univ. Tokyo 8 (1960), 507 - 576.

The University of New South Wales.

ONE UNIVERSE AS A FOUNDATION FOR CATEGORY THEORY

by Saunders Mac Lane

Received May 7, 1969

The development of category theory has posed problems for the set theoretic foundations of Mathematics. These problems arise in the use of collections such as the category of _all_ sets, of _all_ groups, or of _all_ topological spaces. It is the intent of category theory that this "all" be taken seriously; on the other hand, the usual axiomatizations of set theory do not allow the formation of collections such as the set of all sets, or the set of all groups, and indeed the formation of these sets is proscribed precisely in order to avoid the standard paradoxes.

Radical proposals have been made to meet the foundational problems. They might be solved if one dropped the traditional idea that all Mathematics can be developed within one system of (axiomatized) set theory. Indeed, Lawvere [4] has suggested that a foundation might be based upon an axiomatization not of sets, but of the category of all categories. This attractive possibility is not yet fully developed, partly because it offers too many variants, such as axioms for the two-dimensional category of all categories, or perhaps for the (three-dimensional) category of all two-categories.

Pending the formulation and development of these (or other) alternative approaches to foundations, there is an immediate problem of somehow providing an orderly explication of the main results of category theory within the accepted language of set theory. This has often been done by using Gödel-Bernays axiomatization of set theory, which provides for both sets and classes and hence for a class (and thus a category) of all sets or of all groups. However, this approach does not allow for the free formation of functor categories. The Grothendieck school has proposed a strengthening of the axioms of set theory by requiring the existence of many universes (definition below), specifically that each set be a member of a universe (see Gabriel [2] or Verdier [5]). This leads to complications attendant upon change of universe.

It turns out that a flexible and effective formulation of the present notions of category theory can be given with a more modest addition to the standard axiomatic set theory: The assumption that there is <u>one</u> universe. This assumption is close to ideas used by Lawvere to relate his foundation to ordinary set theory, and also to Isbell's use of a single ∞ in [3]. We turn to the details.

Assume first the Zermelo-Fraenkel (ZFC) axioms for set theory, stated in terms of the usual primitive notion $\in$ for membership. These axioms are: Extensionality, the existence of the empty set $\emptyset$ and of the set $\{x,y\}$ for any given sets x,y, the existence to each set x of the union and the power set;

$$\bigcup x = \{t \mid t \in\in x\}, \quad Px = \{b \mid b \subset x\},$$

the axiom of infinity, the axiom of regularity (no infinite descending chain ... $x_n \in x_{n-1} \in \dots \in x_1 \in x_o$), the axiom scheme for replacement and the axiom of choice. With these axioms we can make the standard von Neumann definitions of the ordinal numbers and in particular of the set ω of all natural numbers. The axiom of infinity is then taken to be the assertion that ω exists. The ordered pair $\langle x,y \rangle$ of two sets is then defined as usual, as $\{\{x\}, \{x,y\}\}$; this leads to the standard definitions of cartesian products $u \times v$, of graphs, and of functions.

Now define a universe to be a set U with the following properties:

(i) U is transitive ($x \in g \in U$ implies $x \in U$),

(ii) $\omega \in U$,

(iii) $v \in U$ implies $Pv \in U$,

(iv) $v \in U$ implies $\bigcup v \in U$,

(v) If $f: x \longrightarrow a$ is a surjective function with $x \in U$ and $a \subset U$, then $a \in U$.

In words, this fifth property states that the image of a set x of U under a function of all of whose values are in U is itself a set of U.

From this definition of a universe one may readily derive a number of other elementary closure properties of U. By (i) and (ii) the natural numbers 0, 1, 2, ... are members of U; hence by (v), $x \in U$ and $y \in U$ imply that $\{x\}$, $\{x,y\}$, and the ordered pair $\langle x,y \rangle$ are all in U. Moreover, for any sets $v, w \in U$ the cartesian product $v \times w$ and the set hom(v,w) of all functions $v \longrightarrow w$ are in U; also $y \subset x \in U$ implies $y \in U$. From (iv) and (v) one deduces

(vi) If $I \in U$ and x_i is an I-indexed family of sets with $x_i \in U$ for all $i \in I$, then the union $\bigcup_i x_i$ is a set in U.

Conversely, this condition (vi) implies (v) and (iv), for condition (vi) with the identity indexing function $v \longrightarrow v$ gives condition (iv) and also gives (v) with $a = \bigcup_i f_i$ for $i \in x$. Indeed, Gabriel's definition [2] of a universe is essentially in terms of conditions (i), (ii), (iii), and (iv), with some other minor variations due to Bourbaki's use of the ordered pair as a primitive notion.

Our proposed foundation is now this: The Zermelo-Fraenkel axioms plus the axiom that there exists a universe U. We call a set x _small_ precisely when it is a member of U ($x \in U$); note especially that this is _not_ the French usage, where a set y is called U-small if there is a bijection $y \longrightarrow x$ with $x \in U$. From the definition of a universe it follows readily that the small sets (with the given membership relation $\in$) themselves satisfy the ZFC axioms for set theory. For that matter, if we take "set" to be small set and "class" to be any subset of the universe U, these sets and classes satisfy the usual Gödel-Bernays axioms. Note that our assumptions are stronger than those of Feferman [1], who has ZF together with a symbol (or set) s, satisfying a certain reflection principle. This principle implies that s has some of the properties of a universe U; however, his s, unlike U, is not itself a model of ZFC. Our intention is that the small sets can serve as the objects of Mathematics, while the other sets, not necessarily small, may be used to describe the various categories and functor categories of these Mathematical objects.

We now sketch the formulation of category theory on the basis of our assumptions (ZFC plus one universe U); we repeat that "set" will mean any set, and "small set" a set which is an element of U. A <u>category</u> C will be defined to be a pair of sets--a set of objects and a set of morphisms, together with the usual data giving a domain and a codomain for each morphism, and a composite for suitable pairs of morphisms, all subject to the usual axioms (associativity and existence of identities). A standard example is the category **Sets** of all small sets; its set of objects is exactly the set U (the set of all small sets), and its set of morphisms is a certain subset of $U \times U \times U$ -- namely the set of all those ordered triples $\langle x,y,f\rangle$ such that f is the graph of a function from $x \in U$ to $y \in U$. One may similarly form the category of all small groups or of all small topological spaces, where a "small group" is, of course, a group whose underlying set is small. However, we cannot form the category of <u>all</u> sets or of <u>all</u> groups. On the other hand, if C and D are any two categories, we can use the usual set-theoretic constructions, valid in ZFC, to construct the set of all functors $D \to C$ and the set of all natural transformations between two such functors. These two sets form the functor category C^D -- for any given categories C and D. Observe that we need not require that the domain ("exponent") category be small (A category is small when both the set of objects and the set of morphisms are small sets).

In any category C we may construct subsets of the set of all morphisms by the usual comprehension axiom scheme, which is a consequence of the ZF-axioms. In particular, for any two objects a and b of C we can construct the set hom(a,b) of all morphisms of C from a to b. Conversely, a category may be described in terms of its hom-sets as a set of objects together with functions assigning to each pair a, b of objects a set hom(a,b) and to each triple a,b,c of objects a composition $hom(b,c) \times hom(a,b) \to hom(a,c)$ which is associative and which has the usual identities. This description of a category is equivalent to the preceding one provided the various hom-sets are disjoint; that is, provided $\langle a,b\rangle \neq \langle c,d\rangle$ implies that hom(a,b) and hom(c,d) are disjoint. Should these conditions fail, one can introduce an isomorphic category-with-hom-sets in which these sets are disjoint.

The rub is that hom is not always a bifunctor to a well defined category of sets. Call a category C <u>locally</u> <u>small</u> when the set hom(a,b) is small for each pair of objects a,b of C. Then for any locally small category C there is a hom-functor hom: $C^{op} \times C \to \underline{\underline{Sets}}$, where C^{op} is the usual opposite category to C. Moreover, the categories of all small sets, of all small groups, and the like are all locally small. However, many categories are not locally small; for example, given locally small categories C and D, the functor category C^D need not be locally small, so need not have a hom-functor to $\underline{\underline{Sets}}$.

The category $\underline{\underline{Sets}}$ is also small complete; here a category C is called <u>small</u>-<u>complete</u> if every functor $I \to C$, for I a small category, has a limit. Moreover, for a locally small category C the usual Yoneda embedding $C \to \underline{\underline{Sets}}^{C^{op}}$ given by $a \mapsto hom(-, a)$ does embed C in a small-complete functor category.

These important constructions can still be carried out for categories which are not locally small by using suitable alternative (and large) categories of sets. For this purpose, we shall embed any set S in a larger set $\bar{S}$ as follows. For any set x let Ex denote the set of all elements of elements of x, and set

$$D\,x = x \cup Px \cup Ex.$$

Now let α run through all the small ordinals (i.e., the ordinals in the universe U); and by recursion define sets $D_\alpha S$

$$D_0 S = S, \qquad D_\alpha S = \bigcup_{\beta<\alpha} D(D_\beta S).$$

Finally, let $\bar{S}$ be the union of the $D_\alpha S$ for all these ordinals α. This construction (which is reminiscent of the usual construction from the null set of sets of ordinal rank α) yields an $\bar{S}$ with many closure properties. Because of the use of E in the definition above, one proves readily that $x,y \in \bar{S}$ imply $\{x\}$, $\{x,y\}$, and $\langle x,y\rangle \in \bar{S}$. It follows that $\bar{S}$ is transitive, is closed under power set, union, and under finite cartesian product. Moreover $x \in \bar{S}$ implies that any subset of x is in $\bar{S}$, and for $x,y \in \bar{S}$ the set of all functions from x to y is in $\bar{S}$. Also $\bar{S}$ contains all small ordinals α, and, in particular, contains ω and all the finite ordinal numbers (the natural numbers). This set $\bar{S}$ need not be a universe (axiom (v), the "replacement" schema, need not hold), but it does have the following completeness property: If I is any small set and $x\colon I \to \bar{S}$ is any function, then the cartesian product $\prod_i x_i$ is a set in $\bar{S}$. The proof is straightforward.

From $\bar{S}$ we can form a category of sets, $\underline{\underline{Sets}}_S$ as follows: The set of objects is $\bar{S}$, and the morphisms are all functions $f: x \longrightarrow y$ where $x,y \in \bar{S}$, with the usual composition. The completeness property noted above for $\bar{S}$ shows that this category has all small products. Since it evidently has equalizers of pairs of maps, the standard categorical proof shows that it is small-complete.

We have thus made any set S an object in a small complete category $\underline{\underline{Sets}}_S$ of sets. Taking for S the set of morphisms of any category C whatever, we then have a small complete category of sets which contains this set of morphisms and hence all hom sets for C. These hom sets thus constitute a functor $C^{op} \times C \longrightarrow \underline{\underline{Sets}}_S$. The Yoneda embedding, the Kan extensions, and other existence theorems of category theory which require small-completeness can be similarly handled. This is the asserted sense in which the assumption of one universe, with the use of a variety of categories of sets, provides an adequate foundation for category theory.

BIBLIOGRAPHY

[1] Feferman, S., Set Theoretical Foundations of Category Theory. This volume.

[2] Gabriel, P., Des Catégories Abéliennes, Bull. Soc. Math. de France 90 (1962) 323-448.

[3] Isbell, John, Structure of Categories. Bull. Am. Math. Soc. 72 (1966) 619-655.

[4] Lawvere, F.W., The Category of all Categories as a Foundation of Mathematics, pp. 1-20 in Proceedings of the Conference on Categorical Algebra, La Jolla, 1965, Springer, Heidelberg and New York, 1966.

[5] Verdier, J.L., Topologies et Faisceaux I-IV, in M. Artin and A. Grothendieck, Cohomologie Étale des Schémas, Séminaire de Géometrie Algébrique 1963-64, Institut des Hautes Études Scientifiques, Bures-sur-Yvette, France, 1965.

SET-THEORETICAL FOUNDATIONS OF CATEGORY THEORY

by Solomon Feferman[1]

(With an Appendix by G. Kreisel)

Received May 8, 1969

1. Introduction. Repeatedly when using the notion of set without a clear idea of what sets are meant, mathematicians have produced senseless or even formally contradictory definitions, for instance in the theory of ordinals (cf. Burali-Forti). For mathematical practice it was sufficient to take it that all sets (including sequences of sets, sets of sets, etc.) to be considered belong to some universe U of sets closed with respect to certain operations. When setting up a formal theory, mention of U was not needed because all quantifiers are tacitly supposed to range over such a U. Though some U was essential for the interpretation or foundations of the theory, it was not essential for mathematical practice because no operations were carried out on U; in other words, U was irrelevant to the organization of mathematics.

Category theory introduced a novel element in mathematical practice in that beside such a tacit universe U, one also had distinctions between small and large categories or, as specifically suggested by Grothendieck [2], different kinds of universes. The grossest formulation of this distinction required the addition of strong axioms of infinity. Whatever the intrinsic plausibility of such axioms, they seem to have nothing to do with the actual

1) Research supported by grants DA-ARO(D)-31-124-G985 and NSF-GP-8764.

requirements of category theory but only with the particular formulation adopted. For example, some questions of transferring results about one universe to another arise which seem difficult but irrelevant.

The principal purpose of this paper is to use a well known schema derived in the system of set theory ZFC[2], the so-called <u>reflection-principle</u>, to formulate current category theory set-theoretically without these axioms of universes. The use of this principle for the present purpose was first suggested by Kreisel [5] (p. 118).

First of all we explicitly introduce a symbol <u>s</u> ('s' for the universe of small sets) and a system ZFC/<u>s</u>. This incorporates a form of the reflection principle applied to <u>s</u> as one of its basic axiom schemata; it expresses for each set-theoretic statement φ that φ holds for members of <u>s</u> just in case it holds for members of (the tacitly understood) U. It is shown in § 2 that ZFC/<u>s</u> is a <u>conservative extension</u> of ZFC. This is proved by means of the usual reflection principle in ZFC essentially in the manner of Montague and Vaught [17]. It follows that all set-theoretic statements established in ZFC/<u>s</u> for members of <u>s</u> also hold for all members of the U understood in ZFC.

It appears that current category theory can be carried out naturally in ZFC/<u>s</u>; this is studied in some detail in § 3. In contrast to U, operations are applied to <u>s</u>; but no 'mathematical'

2) The system of Zermelo-Fraenkel with the Axiom of Choice.

assumptions are made on s, only the blanket 'logical' assumption of the reflection principle for s. We have here a typical use of logic in that (i) the mere existence of such an s is used, not detailed properties of it and (ii) it is essential that requirements on s are formulated in a certain language.

Some open problems suggested by the present work are stated in §4, while §5 indicates further work of a different nature (cf. below). It has also seemed useful to add two Appendices. The proposal of Eilenberg-Mac Lane for formulating category theory in the language of the Bernays-Gödel theory of classes is considered briefly in Appendix I (§6). Appendix II (§7), by G. Kreisel, discusses category theory and the foundations of mathematics; this involves some points which are not familiar to the general mathematical public.

In a sequel to this paper a more sophisticated proposal, also due to Kreisel, will be developed. Instead of relying on a blanket principle, one tries to analyze the closure conditions actually needed for current theories on categories of sets and particularly on functor categories (§5.1). Put differently, having substantiated the hypothesis advanced in the present paper that the practice of category theory can be reduced to the principles of ZFC of set theory, we shall go on to an 'algebraization' of the theory of the category of sets, where the closure conditions are formulated algebraically, that is without the use of logically complicated formulae.[3)]

[3)] The proposal is explained in more detail in unpublished notes by Kreisel for a Seminar on Abstract Structures held at Stanford, Fall 1968. I wish to thank Kreisel for contributing Appendix II and for a number of useful comments on a draft of this paper, especially in connection with the Introduction above.

2. The systems of set theory employed and their interrelationships.

2.1. Review of ZF.[4] The language $\mathcal{L}$ of usual formal set theory is given by $\in$, $=$, variables a, b, c, ..., x, y, z, ..., connectives $\sim$ (not), $\wedge$ (and), $\vee$ (or, $\rightarrow$ (implies), $\leftrightarrow$ (iff), and quantifiers $\forall$ (for all) and $\exists$ (there exists). φ, ψ, χ, θ range over (well-formed) formulas of $\mathcal{L}$. The notation $\varphi(x_1, \ldots, x_n)$ only indicates that $x_1, \ldots, x_n$ are among the free variables of φ. A sentence is a formula without free variables.

ZF consists of the axioms of (i) extensionality, (ii) empty set, (iii) (unordered) pairs, (iv) union set, (v) power set, (vi) infinity, (vii) foundation, and (viii) the instances of the replacement axiom schema, i.e., for each formula $\varphi(x,y)$ with free variables $x,y,a,z_1, \ldots, z_n$ altogether:

$$\forall a,z_1,\ldots,z_n\{\forall x,y_1,y_2[\varphi(x,y_1)\wedge\varphi(x,y_2)\rightarrow y_1=y_2]\rightarrow$$
$$\exists b\forall y[y\in b\leftrightarrow\exists x(x\in a\wedge\varphi(x,y))]\ .$$

An immediate consequence of (viii) for any formula $\varphi(x)$ is the separation axiom

$$\text{(viii)'}\quad \forall a,z_1,\ldots,z_n\exists b\forall x[x\in b\leftrightarrow x\in a\wedge\varphi(x)]\ .$$

The axioms Z of Zermelo consist of (i)-(vii) and all instances of (viii)'.

---

4) The recent text of Krivine [8] gives a very readable, informative, yet compact exposition of Zermelo-Fraenkel set theory including the reflection principles used below (as well as much else of general interest).

If it is proved for a formula $\Psi(x)$ that (for any choice of its other free variables) $\exists b \forall x[x \in b \leftrightarrow \Psi(x)]$, this unique b is denoted by $\{x|\Psi(x)]$. In particular, we can always intoduce $\{x|x \in a \wedge \varphi(x)]$, and we can introduce $\{y|\exists x(x \in a \wedge \varphi(x,y)]$ for those $\varphi(x,y)$ such that $\forall x, y_1, y_2[\varphi(x,y) \wedge \varphi(x,y_2) \rightarrow y_1 = y_2]$ has been proved. By (ii)-(v) we can introduce $0 = \{x|x \neq x\}$, $\{a,b\} = \{x|x = a \vee x = b\}$, $\{a\} = \{a,a\}$, $\bigcup a = \{y|(\exists x)(y \in x \wedge x \in a)]$, and $P(a) = \{x|(\forall y)(y \in x \rightarrow y \in a)\} = \{x|x \subseteq a\}$. As usual one defines $a \cup b = \bigcup \{a,b\}$, $a \cap b = \{x|x \in a \wedge x \in b\}$, $a - b = \{x|x \in a \wedge x \notin b\}$, $a' = a \cup \{a\}$, $\langle a,b \rangle = \{\{a\}, \{a,b\}\}$, and $a \times b = \{z|\exists x(x \in a \wedge y \in b \wedge z = \langle x,y \rangle)\}$. Fn(f) expresses that f is a function, i.e., is a set of ordered pairs such that $\forall x, y_1 y_2[\langle x,y_1 \rangle \in f \wedge \langle x,y_2 \rangle \in f \rightarrow y_1 = y_2]$. $\mathfrak{D}(f) = \{x|\exists y \langle x,y \rangle \in f)\}$ and $\mathfrak{R}(f) = \{y|\exists x(\langle x,y \rangle \in f)]$. f(x) denotes the unique y such that $\langle x,y \rangle \in f$ when Fn(f). $\bigcup_{x \in a} f(x) = \bigcup \{y|\exists x(x \in a \wedge y = f(x)]$. The Cartesian product $\prod_{x \in a} f(x)$ is defined as $\{g|Fn(g) \wedge \mathfrak{D}(g) = a \wedge \forall x(x \in a \rightarrow g(x) \in f(x))\}$;then $\prod_{x \in a} f(x) \subseteq P(a \times \bigcup_{x \in a} f(x))$. Let $\prod a = \prod_{x \in a} x$; the statement $\forall a(0 \notin a \rightarrow \prod a \neq 0)$ is equivalent to the axiom of choice, AC.

The axiom of foundation expresses that $\in$ is well-founded. The ordinals can be identified with transitive sets $\underline{a}$ (i.e., such that $\bigcup a \subseteq a$) which are linearly ordered by $\in$. Ord(a) expresses that x is an ordinal. α, β, γ, ... range over ordinals in the following. The successor $\alpha + 1$ of any ordinal α is α'. The axiom of infinity gives the existence of a limit ordinal α, i.e., one such that $0 \in \alpha \wedge \forall x(x \in \alpha \rightarrow x' \in \alpha)$; the least such ordinal

ω is the set of natural numbers. The restriction of $\in$ to ordinals is denoted by $<$.

By the axiom of replacement one can prove that for each γ there is a unique f such that Fn(f) and $\mathcal{D}(f) = \gamma$ and for any $\alpha < \gamma$, $f(\alpha) = \bigcup_{\xi<\alpha}(f(\xi) \cup P(f(\xi)))$. R_α is $f(\alpha)$ for any such f.[5] It follows from foundation that $\forall x \exists \alpha (x \in R_\alpha)$. The least such α must be a successor, so for each x there is a unique α with $x \in R_{\alpha+1} - R_\alpha$; this α is denoted by rank (x). Given any $\varphi(x)$ it follows that there is a set of all x of least rank satisfying $\varphi(x)$, i.e.,
$\{x \mid \varphi(x) \wedge \sim \exists\, y(\varphi(y) \wedge \mathrm{rank}(y) < \mathrm{rank}(x))\}$.

2.2 <u>The system</u> ZF/<u>s</u>.

Let <u>s</u> be a constant symbol and let $\mathcal{L}_{\underline{s}}$ be the language $\mathcal{L}$ with <u>s</u> added. Write $\exists x \in \underline{s}\varphi(x)$ and $\forall x \in \underline{s}\varphi(x)$ for $\exists x[x \in \underline{s} \wedge \varphi(x)]$ and $\forall x[x \in \underline{s} \rightarrow \varphi(x)]$, resp. For any formula φ, $\varphi^{(\underline{s})}$ denotes the result of relativizing all quantifiers to φ, i.e., each $(\exists x)$ is replaced by $(\exists x \in \underline{s})$ and $(\forall x)$ by $(\forall x \in \underline{s})$.

The axioms of ZF/<u>s</u> are taken to consist of the following in the basic symbolism of $\mathcal{L}_{\underline{s}}$.

(1) the axioms of ZF

(2) $\exists x(x \in \underline{s})$

(3) $\forall x,y[y \in x \wedge x \in \underline{s} \rightarrow y \in \underline{s}]$

---

[5] In other words, R_α is the result of iterating α times the <u>cumulative power set operation</u> R: $a \mapsto a \cup P(a)$. The sequence of R_α's is called the <u>cumulative hierarchy</u>; cf. § 7.1.

(4) $\forall x,y[x \in \underline{s} \wedge \forall z(z \in y \rightarrow z \in x) \rightarrow y \in \underline{s}]$

(5) $\forall x_1 \in \underline{s} \ldots \forall x_n \in \underline{s}[\varphi^{(\underline{s})}(x_1,\ldots,x_n) \leftrightarrow \varphi(x_1,\ldots,x_n)]$

for each formula φ of $\mathcal{L}$ with free variables $x_1, \ldots, x_n$.

With the abbreviations introduced above, (2) $\underline{s} \neq 0$, (3) $\bigcup \underline{s} \subseteq \underline{s}$, (4) $x \in \underline{s} \wedge y \subseteq x \rightarrow y \in \underline{s}$. (5) is the <u>reflection principle</u>. Model-theoretically this expresses that if (M,E,s) is any model of ZF/$\underline{s}$ and $M_s = \{x | x \in M$ and $xEs\}$ and xE_sy iff $x, y \in M_s$ and xEy, then (M,E) is an <u>elementary extension</u> of (M_s,E_s), i.e., these satisfy the same formulas of $\mathcal{L}$.

It follows from (1) and (5) that if φ is any sentence of $\mathcal{L}$ and ZF $\vdash \varphi$ (φ is derivable from ZF) then ZF/$\underline{s} \vdash \varphi^{(\underline{s})}$. Each of the axioms (ii)-(v) of ZF is of the form $\forall a_1 \ldots \forall a_k \exists b \forall x[x \in b \leftrightarrow \psi(x,a_1, \ldots, a_k)]$. The relativization of this is

$$\forall a_1 \in \underline{s} \ldots \forall a_k \in \underline{s} \exists b \in \underline{s} \forall x \in \underline{s}[x \in b \leftrightarrow \psi^{(\underline{s})}(x,a_1, \ldots, a_k)]$$

By transitivity of $\underline{s}$ it follows that $0 \in \underline{s}$ and if $a, a_1, a_2 \in \underline{s}$ then $\{a_1,a_2\} \in \underline{s}$ and $\bigcup a \in \underline{s}$. But also $[a \in \underline{s} \rightarrow P(a) \in \underline{s}]$ by (4). Then the closure of $\underline{s}$ under all the other operations defined above is quite direct. The relativization to $\underline{s}$ of the statement φ that there is an ordinal containing 0 and closed under ' is equivalent to φ; hence $\omega \in \underline{s}$.

The relativization of the replacement axioms (viii) are as follows, for any formula $\varphi(x,y)$ of $\mathcal{L}$ with free variables $x,y,a,z_1, \ldots, z_n$ altogether:

$$\forall a, z_1, \ldots, z_n \in \underline{s}\{\forall x,y_1,y_2 \in \underline{s}[\varphi^{(\underline{s})}(x,y)\wedge\varphi^{(\underline{s})}(x,y_2) \rightarrow y_1 = y_2]$$

$$\rightarrow \exists b \in \underline{s}\forall y \in \underline{s}[y \in \underline{b} \leftrightarrow \exists x(x \in a \wedge \varphi^{(\underline{s})}(x,y))]\} .$$

By the reflection principle, this implies

$$(\mathrm{viii})_{\underline{s}} \quad \forall a,z_1, \ldots, z_n \in \underline{s}\{\forall x,y_1,y_2[\varphi(x,y_1)\wedge\varphi(x,y_2) \rightarrow y_1 = y_2]$$

$$\rightarrow \exists b \in \underline{s}\forall y[y \in \underline{b} \leftrightarrow y \in \underline{s} \wedge \exists x \in a\varphi(x,y)]\} .^{6)}$$

Inspection of the instances of replacement used in the proof of the existence of the R_α together with closure of $\underline{s}$ under P gives:

$$\forall \alpha \in \underline{s}(R_\alpha \in \underline{s}) .$$

Let $\sigma = \{x|x \in \underline{s} \wedge \mathrm{Ord}(x)\}$. Then σ is an ordinal and is the least one not in $\underline{s}$. Moreover, σ is a limit ordinal and $\sigma > \omega$. Finally, $\underline{s} = R_\sigma$.

The statements $(\mathrm{viii})_{\underline{s}}$ may be considered to express the inaccessibility of s under all functions definable in $\mathcal{L}$.

The following abbreviations will facilitate comparison with the standard notion of inaccessibility. Let R(x) be $\exists\alpha(\alpha > \omega \wedge \text{limit }(\alpha) \wedge x = R_\alpha)$, and let In(x) be $\forall f(\mathrm{Fn}(f) \wedge \mathfrak{D}(f) \in x \wedge \mathcal{R}(f) \subseteq x \rightarrow \mathcal{R}(f) \in x)$; finally let RIn(x) be $R(x) \wedge \mathrm{In}(x)$. Then RIn(x) holds iff $\alpha = \mathrm{rank}(x)$ is a strongly inaccessible cardinal in the usual sense (assuming AC).

---

6) The restriction here that φ not contain $\underline{s}$ is important in the proof of the basic result in the next section. This restriction is essential in Axiom (5), otherwise we could conclude $\exists x \in \underline{s}(\underline{s} \in x)$ from $\exists x(\underline{s} \in x)$.

Let $\mathrm{In}_{\mathcal{L}}(\underline{s})$ be the set of statements $(\mathrm{viii})_{\underline{s}}$, and let $\mathrm{RIn}_{\mathcal{L}}(\underline{s})$ consist in addition of $R(\underline{s})$. It is clear from the separation axioms that

$$ZF + \mathrm{RIn}(\underline{s}) \vdash \mathrm{RIn}_{\mathcal{L}}(\underline{s}) .$$

On the other hand, the argument above shows that also

$$ZF/\underline{s} \vdash \mathrm{RIn}_{\mathcal{L}}(\underline{s}) .$$

The theory $ZF(\underline{s}) = ZF + \mathrm{RIn}_{\mathcal{L}}(\underline{s})$ is thus a common subtheory of $ZF + \mathrm{RIn}(\underline{s})$ and of $ZF/\underline{s}$. The formalization of category theory in set theory will be studied in $ZF(\underline{s})$ in §3.1-3.3. The reflection principle will then be adjoined to yield further consequences.

The basic result of the next section does not assume the axiom of choice, but also holds when it is added.[7] Let $ZFC = ZF + AC$, $ZFC/\underline{s} + AC$. Assuming $ZFC(\underline{s})$, note that

$$\forall a \in \underline{s}[0 \notin a \rightarrow \exists f(\mathrm{Fn}(f) \wedge f \in \underline{s} \wedge \forall x \in a(f(x) \in x))] ,$$

i.e., there is always a choice function in $\underline{s}$ for sets in $\underline{s}$. Then every $a \in \underline{s}$ is in 1-1 correspondence by a mapping in $\underline{s}$ with an ordinal $\alpha \in \underline{s}$, so $a \in \underline{s} \rightarrow \mathrm{card}(a) \in \underline{s}$ (identifying cardinals with initial ordinals). Since $\aleph_\alpha \leq \mathrm{card}(R_{\omega+\alpha})$ it follows that $\alpha < \sigma \rightarrow \aleph_\alpha < \sigma$. In particular, $\aleph_0, \aleph_1, \ldots, \aleph_{\aleph_0}, \aleph_{\aleph_1}, \ldots, \aleph_{\aleph_{\aleph_0}}$ can all be proved to belong to $\underline{s}$ in the system $ZFC(\underline{s})$.

---

[7] We follow the conventional separation of AC from ZF. However, there is no reason to treat AC on a separate footing if one accepts the basic interpretation of set theory in the cumulative hierarchy (§7.1).

2.3. <u>The basic metamathematical result.</u>

<u>Theorem 1. ZF/s is a conservative extension of</u> ZF.

<u>Proof</u>. Suppose Ψ is a sentence of $\mathcal{L}$ and ZF/$\underline{s} \vdash \Psi$. It is to be shown that ZF $\vdash \Psi$. Ψ is logically derivable from a finite subset T of ZF/$\underline{s}$. T can be put in the form $\{\theta_1(\underline{s}), \ldots, \theta_m(s)\}$ where each $\theta_i(a)$ is a formula of $\mathcal{L}$. It is sufficient to show that

$$\text{ZF} \vdash \exists a[\theta_1(a) \wedge \ldots \wedge \theta_m(a)] .$$

Finitely many instances of ZF extended to $\mathcal{L}_{\underline{s}}$ occur among the $\theta_i(\underline{s})$; in these cases $\forall a \theta_i(a)$ is provable in ZF. Any $a = R_\alpha$ satisfies (2)-(4) of ZF/$\underline{s}$. Finitely many instances of the reflection principle accur among the $\theta_i(\underline{s})$, say for formulas $\varphi_j(x_1, \ldots, x_n)(j = 1, \ldots, k)$. Hence it is sufficient to prove in ZF that for some ordinal α and for $a = R_\alpha$,

$$\forall x_1 \in a \ldots \forall x_n \in a[\varphi_j^{(a)}(x_1, \ldots, x_n) \leftrightarrow \varphi_j(x_1, \ldots, x_n)] .$$

This is by formalization of the argument in [17] by which if R_{α_0} is any model of ZF then there exists $\alpha \leq \alpha_0$ such that α is accessible and $(R_{\alpha_0}, \in \upharpoonright R_{\alpha_0})$ is an elementary extension of $(R_\alpha, \in \upharpoonright R_\alpha)$. This makes use of a modified Skolem-Löwenheim procedure, as follows.

Each of the φ_j can be assumed to be in prenex form, say $Q_1 u_1 \ldots Q_r u_r \chi_j(x_1, \ldots, x_n, u_1, \ldots, u_r)$ with each $Q_i = \exists$ or $\forall$. Let $\varphi_{i,j}(x_1, \ldots, x_n, u_1, \ldots, u_i)$ be $Q_{i+1} u_{i+1} \ldots Q_r u_r \chi_j(x_1, \ldots, x_n, u_1, \ldots, u_r)$. When Q_i is $\exists$ we have a formally definable function

$_{,j}(x_1,\ldots,x_n, u_1,\ldots,u_{i-1})$

$= \{u_i | u_i$ is of least rank such that $\varphi_{i,j}(x_1,\ldots,x_n,u_1,\ldots,u_{i-1}, u_i)\}$.

en

$_i \varphi_{i,j}(x_1,\ldots,x_n, u_1,\ldots,u_{i-1}, u_i) \leftrightarrow$

$\exists u_i[u_i \in F_{i,j}(x_1,\ldots,x_n, u_1,\ldots,u_{i-1}) \wedge \varphi_{i,j}(x_1,\ldots,x_n, u_1,\ldots,u_{i-1}, u_i)]$.

a dual definition when Q_i is $\forall$ we get $F_{i,j}$ such that

$_i \varphi_{i,j}(x_1,\ldots,x_n, u_1,\ldots,u_{i-1}, u_i) \leftrightarrow$

$\forall u_i[u_i \in F_{i,j}(x_1,\ldots,x_n, u_1,\ldots,u_{i-1}) \rightarrow \varphi_{i,j}(x_1,\ldots,x_n, u_1,\ldots,u_{i-1},u_i)]$.

e $F_{i,j}$ act like Skolem functions for the given formulas. Given

b, let

$= b \cup \bigcup F_{i,j}(x_1,\ldots,x_n, u_1,\ldots,u_{i-1})$

$[x_1,\ldots,x_n, u_1,\ldots,u_{i-1} \in b,\ 1 \leq j \leq k,\ 1 \leq i \leq r]$.

s exists by the axiom of replacement. For any β let $\beta^* = \text{rank}(R^*_\beta)$;

en $R^*_\beta \subseteq R_{\beta^*}$. Finally, take $\alpha = \bigcup_{n<\omega} \beta_n$ where $\beta_0 = 0$,

$_{n+1} = \beta^*_n$, and let $a = R_\alpha$. It follows by induction on $r - i$,

$\leq i \leq r$, that for any $x_1,\ldots,x_n, u_1,\ldots,u_i \in a$,

$^{(a)}_{i,j}(x_1,\ldots,x_n, u_1,\ldots,u_i) \leftrightarrow \varphi_{i,j}(x_1,\ldots,x_n, u_1,\ldots,u_i)$.

tting $i = 0$ completes the proof.

This result generalizes immediately to any extension of

/s by sentences of $\mathcal{L}$.

Corollary. Suppose T is any set of sentences of $\mathcal{L}$. Then ZF/$\underline{s}$ + T is a conservative extension of ZF + T. In particular, ZFC/$\underline{s}$ is a conservative extension of ZFC.

Proof. If Ψ is a sentence of $\mathcal{L}$ and $(ZF/\underline{s} + T) \vdash \Psi$ then $ZF/\underline{s} \vdash (\Theta_1 \wedge \ldots \wedge \Theta_m \rightarrow \Psi)$ for some $\Theta_1, \ldots, \Theta_m \in T$. Then $ZF \vdash (\Theta_1 \wedge \ldots \wedge \Theta_m \rightarrow \Psi)$ by Theorem 1, so $(ZF + T) \vdash \Psi$.

This corollary does not give us information about extensions of ZF/$\underline{s}$ by sentences containing $\underline{s}$. In particular, it is of interest to consider the adjunction of RIn($\underline{s}$). Note that $ZF/\underline{s} + RIn(\underline{s}) \vdash In_\infty$, where In_∞ is the statement $\forall a\, \exists b[RIn(b) \wedge a \in b]$. For, we can derive: (i) $\forall a \in \underline{s}\, \exists b[RIn(b) \wedge a \in b]$, (ii) $\forall a \in \underline{s}[\exists b \in \underline{s}(RIn^{(\underline{s})}(b) \wedge a \in b) \leftrightarrow \exists b(RIn(b) \wedge a \in b)]$, and (iii) $\forall a \in \underline{s}\, \forall b \in \underline{s}[RIn^{(\underline{s})}(b) \leftrightarrow RIn(b)]$. Thus the set of sentences of $\mathcal{L}$ derivable from ZF/$\underline{s}$ + RIn($\underline{s}$) includes strong existential statements concerning inaccessibles. A direct axiomatization of this set has been given by Lévy [11], Theorem 1: Let ZM[8)] be ZF together with a schema which expresses that each normal definable function of ordinals has arbitrarily large inaccessible fixed points; then ZF/$\underline{s}$ + RIn($\underline{s}$) is a conservative extension of ZM. The proof is by an extension of the arguments of [17] as given above for Theorem 1. Then ZFC/$\underline{s}$ + RIn($\underline{s}$) is a conservative extension of ZMC.

It is conceivable that some applications to category theory would essentially require the use of a succession of sets $\underline{s}_o, \underline{s}_1, \ldots$

---

8) 'M' for 'Mahlo'.

having enough properties of "universes". While this seems unlikely, it is easy to obtain appropriate conservative extension results, as follows. Let $\mathcal{L}_{\underline{s}_0,\ldots,\underline{s}_k,\ldots}$ be obtained from $\mathcal{L}$ by adjoining distinct constant symbols $\underline{s}_0,\ldots,\underline{s}_k,\ldots(k<\omega)$. Let ZF/$\underline{s}_0,\ldots,\underline{s}_k,\ldots$ consist of the axioms of ZF extended to $\mathcal{L}_{\underline{s}_0,\ldots,\underline{s}_k,\ldots}$, the axioms (2)-(5) with $\underline{s}_k$ substituted for $\underline{s}$, and the axioms ($\underline{s}_k \in \underline{s}_{k+1}$), for each k. <u>Then each of the conservative extension results proved above also holds if</u> 'ZF/$\underline{s}$' <u>is replaced</u> by 'ZF/$\underline{s}_0,\ldots,\underline{s}_k,\ldots$'. This is proved by a simple modification of the above arguments.

3. <u>Formulation of category theory in</u> ZFC/$\underline{s}$.

3.1. <u>'$\underline{s}$' for 'small'</u>. We now consider the formulation of the basic notions of category theory in the language $\mathcal{L}_{\underline{s}}$. This is done more or less informally, assuming to begin with only the principles expressed in ZFC + $\mathrm{RIn}_{\mathcal{L}}(\underline{s})$. All objects dealt with by these principles are called <u>sets</u> or <u>collections</u> (with the same meaning.) To facilitate matters, we also expand the letters which may be used as variables ranging over sets, including upper case latin letters A,B,C,...,X,Y,Z, script letters $\mathcal{A},\mathcal{B},\mathcal{C},\ldots$ and Greek letters not already used in §3.

$\mathcal{A}$ is called a <u>category</u>, and we write Cat($\mathcal{A}$), if $\mathcal{A}$ is an ordered pair (M,C) where C is a function with $\mathcal{D}(C) \subseteq M^2$, $\mathcal{R}(C) \subseteq M$, satisfying the usual conditions (e.g. [1], p.5) when M is the set of <u>morphisms</u> of $\mathcal{A}$ and C is the (partial) <u>composition</u> operation on morphisms of $\mathcal{A}$. We write $M = M_{\mathcal{A}}$, $C = C_{\mathcal{A}}$. The identity morphisms and the objects of $\mathcal{A}$ are then defined as usual; the set $Ob_{\mathcal{A}}$ of objects of $\mathcal{A}$ is a subset of $M_{\mathcal{A}}$.

A set X is said to be small if $X \in \underline{s}$. Then a category $\mathcal{A}$ is small iff $M_{\mathcal{A}}$ and $C_{\mathcal{A}}$ are in $\underline{s}$. $\mathcal{A}$ is said to be locally small [12] if for each $A, B \in Ob_{\mathcal{A}}$, the set $Hom_{\mathcal{A}}(A,B) = \{f | f \in M_{\mathcal{A}}$ and $f: A \to B\} \in \underline{s}$. Sometimes this condition is taken as part of the definition of being a category, e.g. in [1], [14], but this is not required here.

It should be clear now what is meant by the category of all small sets, the category of all small Abelian groups, etc. These notions are readily generalized as follows. By a morphism between sets is meant a triple $f = (f_1,A,B)$ where f_1 is a function with $\mathfrak{D}(f_1) = A$ and $\mathfrak{R}(f_1) \subseteq B$; one writes $f: A \to B$ in this case. Let $Hom(A,B) = \{f | f: A \to B\}$. If $f: A \to B$ and $g: A' \to B'$ are morphisms between sets, the composition gf is defined and is h iff $B = A'$ and $h = (h_1,A,B')$ with $h_1(x) = g_1(f_1(x))$ for each $x \in A$. By a category of sets is meant any category $\mathcal{A}$ where $M_{\mathcal{A}}$ is a collection of morphisms between sets and $C_{\mathcal{A}}(g,f) = gf$ for any $f,g \in M_{\mathcal{A}}$ such that gf is defined and $gf \in M_{\mathcal{A}}$. $\mathcal{A}$ is said to be full if for any $A,B \in Ob_{\mathcal{A}}$, $Hom_{\mathcal{A}}(A,B) = Hom(A,B)$. In this case, whenever $f,g \in M_{\mathcal{A}}$ and gf is defined we have $gf \in M_{\mathcal{A}}$.[9)]

For any collections K_o, K_1 of sets, let $\mathcal{S}_{K_o,K_1}$ consist of all morphisms (f_1,A,B) such that $A,B \in K_o$ and $f_1 \in K_1$, together

9) One will only be interested in categories of sets which satisfy this closure condition. However, assumption of this would complicate the explanation of $\mathcal{S}_{K_o,K_1}$ to be considered next.

with the restriction of composition to these morphisms. Then $\mathcal{S}_{K_o,K_1}$ is a category of sets, called the <u>category of all sets in</u> K_o <u>with morphisms in</u> K_1. For any collection K, let $\mathcal{S}_K = \mathcal{S}_{K,K}$; this is called the <u>category of all sets in</u> K. If K is closed under Cartesian product and subsets then $\mathcal{S}_K$ is full; for, if $A,B \in K$ and f_1 is a function with $\mathcal{D}(f_1) = A, \mathcal{R}(f_1) \subseteq B$ we have $f_1 \subseteq A \times B$, so $f_1 \in K$. With this notation, the category of all small sets is just $\mathcal{S}_{\underline{s}}$. By the preceding remark, $\mathcal{S}_{\underline{s}}$ is full and locally small.

One defines similarly: (i) a category of Abelian groups, (ii) the category $\mathcal{AG}_{K_o,K_1}$ of all Abelian groups in K_o with (group) morphisms in K_1, and (iii) the category $\mathcal{AG}_K = \mathcal{AG}_{K,K}$ of all Abelian groups and morphisms in K. The category of all small Abelian groups is just $\mathcal{AG}_{\underline{s}}$, which is full and locally small.

The notions of (covariant) <u>functor</u> $F: \mathcal{A} \to \mathcal{B}$ between two categories $\mathcal{A}$, $\mathcal{B}$, and of <u>natural transformation</u> $\eta: F \to G$ between two such functors are defined in the usual way. When considering fixed categories $\mathcal{A}$, $\mathcal{B}$, functors can simply be regarded as certain functions F with $\mathcal{D}(F) = M_{\mathcal{A}}$, $\mathcal{R}(F) \subseteq M_{\mathcal{B}}$. (This must be modified when dealing with functors between members of an arbitrary collection of categories; cf. below.) In any case, natural transformations $\eta: F \to G$ are taken to be triples $\eta = (\eta_1, F, G)$ such that η_1 is a function with $\mathcal{D}(\eta_1) = Ob_{\mathcal{A}}$, $\mathcal{R}(\eta_1) \subseteq M_{\mathcal{B}}$ satisfying the usual commutativity condition. The set $\{\eta \mid \eta : F \to G\}$ of all such transformations is denoted by Nat(F,G). The category $\mathcal{B}^{\mathcal{A}}$ <u>of all functors from</u> $\mathcal{A}$ <u>to</u> $\mathcal{B}$ has as morphisms all nat. transfs.

between functors F: $\mathcal{A} \to \mathcal{B}$, with the usual composition $\eta\zeta$.

The type of relativization of categories of sets, Abelian groups, etc., considered above is more or less familiar. Curiously, it seems not to have been extended to categories of functors, where it is just as appropriate. By <u>a category</u> $\mathcal{F}$ <u>of functors from</u> $\mathcal{A}$ <u>to</u> $\mathcal{B}$ is meant a subcategory of $\mathcal{B}^{\mathcal{A}}$. In this case $Ob_{\mathcal{F}}$ is a collection of functors from $\mathcal{A}$ to $\mathcal{B}$, and for each F, G $\in Ob_{\mathcal{F}}$, the set of morphisms in $\mathcal{F}$ from F to G is a subset $Nat_{\mathcal{F}}(F,G)$ of Nat(F,G); $C_{\mathcal{F}}(\eta,\zeta) = \eta\zeta$ when $\eta\zeta$ is defined and is a morphism of $\mathcal{F}$. $\mathcal{F}$ is said to be <u>full</u> if $Nat_{\mathcal{F}}(F,G) = Nat(F,G)$ for any F, G $\in Ob_{\mathcal{F}}$.

For any collections K_o, K_1 of sets, let $(\mathcal{B}^{\mathcal{A}})_{K_o,K_1}$ have as morphisms all nat. transfs. $\eta = (\eta_1,F,G)$ such that F, G $\in K_o$ and $\eta_1 \in K_1$, together with the restriction of the usual composition to these morphisms. Then $(\mathcal{B}^{\mathcal{A}})_{K_o,K_1}$ is a category of functors from $\mathcal{A}$ to $\mathcal{B}$, called the <u>category of all functors in</u> K_o <u>with nat. transfs. in</u> K_1. For any collection K, let $(\mathcal{B}^{\mathcal{A}})_K = (\mathcal{B}^{\mathcal{A}})_{K,K}$; this is called the <u>category of all functors in</u> K.

A case of special interest is $(\mathcal{B}^{\mathcal{A}})_{\underline{s}}$ when $\mathcal{A}$ is small and $\mathcal{B}$ is locally small. In this case $(\mathcal{B}^{\mathcal{A}})_{\underline{s}}$ is full and locally small. For, given any functors F, G $\in (\mathcal{B}^{\mathcal{A}})_{\underline{s}}$, we have F, G $\in \underline{s}$ so that $\mathcal{R}(F)$, $\mathcal{R}(G)$ are small. Any nat. transf. $\eta = (\eta_1,F,G)$ has $\mathcal{D}(\eta_1) = Ob_{\mathcal{A}}$ and $\eta_1(A) \in Hom_{\mathcal{B}}(F(A),G(A))$ for each $A \in \mathcal{A}$. But then $\eta_1 \subseteq Ob_{\mathcal{A}} \times \bigcup Hom(B,C)[B \in \mathcal{R}(F),\ C \in \mathcal{R}(G)] \in \underline{s}$, so $\eta_1 \in \underline{s}$; this also shows Nat(F,G) $\in \underline{s}$. This argument can be analyzed to give some general sufficient conditions for $(\mathcal{B}^{\mathcal{A}})_K$ to be full.

Functors between members of a collection L of categories may be identified with triples $F = (F_1, \mathcal{A}, \mathcal{B})$ where $\mathcal{A}, \mathcal{B} \in L$ and F_1 is a function from $M_{\mathcal{A}}$ with $\mathcal{R}(F_1) \subseteq M_{\mathcal{B}}$. However, this may have the effect of making F "large" even where F_1 is not large; this effect is multiplied when forming nat. transfs. (η_1, F, G). A simple alternative is available when L is an <u>indexed collection</u>, $L = \{\mathcal{A}_i | i \in I\}$. In this case we identify functors $F: \mathcal{A}_i \rightarrow \mathcal{A}_j$ with triples (F_1, i, j). It turns out that a systematic use of indexed collections, for categories of sets as well as functors, greatly facilitates a development of category theory under everyday set-theoretical assumptions; this will be described in the sequel to this paper.

3.2. <u>Case study of current category theory: two hypotheses</u>. With the preceding explanations, every notion of current category theory can be translated into set-theoretical terms, taking membership in $\underline{s}$ to express being small. Suppose φ is a sentence in $\mathcal{L}_{\underline{s}}$ which is the translation of a result in category theory; in this case we shall call φ a theorem of category theory.

<u>Hypothesis I.</u> <u>Every theorem of current category theory is provable</u> <u>in</u> ZFC + RIn($\underline{s}$).

The evidence for this seems to be conclusive. It is easy to go over a large number of results of current category theory, checking that RIn($\underline{s}$) implies all that is used about the notion of smallness. In another form, this hypothesis has been advanced by a number of workers in the field such as Lawvere [10], Mac Lane [16]: only one inaccessible (or universe) is needed for category theory. (Cf. also §6.)

ZFC + RIn($\underline{s}$) does not contain the reflection principle for $\underline{s}$. This principle is very important if one wishes to see how to extend, automatically, results about small sets (in particular, small structures and categories) to arbitrary sets (cf. §3.4 below). However, as we have seen in §2.3, ZFC/$\underline{s}$ + RIn($\underline{s}$) is extremely strong. We wish then to see to what extent ZFC/$\underline{s}$ + $RIn_{\mathcal{L}}(\underline{s})$ is adequate to category theory. Without question there are certain results which, as currently stated, definitely require the hypothesis embodied in RIn($\underline{s}$) that for any function f considered with $\mathcal{D}(f) \in \underline{s}$ and $\mathcal{R}(f) \subseteq \underline{s}$, we have $\mathcal{R}(f) \in \underline{s}$. (One such result will be taken up in the next section.) However, it seems that the only applications which are made of such results are for set-theoretically definable functions, i.e., for which we have a formula $\varphi(x,y)$ with $f(x) = y \leftrightarrow \varphi(x,y)$. In these cases, the required condition $\mathcal{R}(f) \in \underline{s}$, is derivable from $In_{\mathcal{L}}(\underline{s})$. If the presumption is correct, each such result has a formally weaker form, which is applied only to $f \in \underline{s}$, but which is fully adequate to the applications. It is in this sense that I propose the following.

Hypothesis II. Every theorem of current category theory is provable, or has an adequate version which is provable in ZFC($\underline{s}$), i.e. ZFC + $RIn_{\mathcal{L}}(\underline{s})$.

3.3. <u>Case study of current category theory: specific illustrations</u>. Three results, said to be representative in their use of conditions of smallness, are examined here: the Adjoint Functor Theorem (AFT), Yoneda's Lemma (YL), and the existence of Kan Extensions (KE).

(i) <u>AFT</u>. Mac Lane [16] has analyzed general closure conditions needed for Freyd's AFT and has incorporated these in an axiomatic theory of "schools". Members of a "normal" subschool are called small. It is thus sufficient to check that his conditions on a school and normal subschool are met when one translates "items" and "classes" to be arbitrary sets (in ZFC($\underline{s}$)) and members of the normal subschool to be the sets in $\underline{s}$. This is readily done, and it follows that the translated theorem is provable in ZFC($\underline{s}$).

A second question is whether the standard applications are derivable in ZFC($\underline{s}$) from this interpretation of Mac Lane's AFT. A typical example is given by the forgetful functor F: $\mathcal{AG}_{\underline{s}} \to \mathcal{S}_{\underline{s}}$. First of all, $\mathcal{AG}_{\underline{s}}$ is small-complete; given any small family $\langle A_i \rangle_{i\in I}$ of objects in $\mathcal{AG}_{\underline{s}}$, i.e., where $\{\langle i, A_i\rangle \mid i \in I\} \in \underline{s}$, we have a product $(\prod_{i\in I} A_i) \in \underline{s}$. The existence of equalizers is trivial. We have seen in §3.1 that $\mathcal{AG}_{\underline{s}}$ is locally small. For each $B \in \underline{s}$, the solution set condition calls for a pair of small families $\langle A_i \rangle_{i\in I}$, $\langle g_i \rangle_{i\in I}$ such that

(a) $g_i\colon B \to F(A_i)$ for each $i \in I$, and

(b) whenever $A \in Ob_{\mathcal{A}}$ and $g\colon B \to F(A)$ then $g = F(f)g_i$ for some $i \in I$ and $f\colon A_i \to A$.

Suppose $B \in R_\beta$ with $\beta < \sigma$. We can take $\langle A_i \rangle_{i\in I}$ to be the collection of all Abelian groups $\langle B_i, +_i\rangle$ with $B_i \in R_{\beta+1}$, and with each B_i repeated as often as there are morphisms $g_i: B \to B_i$. Then $\langle A_i \rangle_{i \in I}$ and $\langle g_i \rangle_{i \in I}$ are in $\underline{s}$, as required.

(ii) <u>YL</u>. Let $\mathcal{A}$ be any locally small category and let $\mathcal{S} = \mathcal{S}_{\underline{s}} =$ the category of small sets. We have for each $A \in \mathcal{A}$ a functor $H_A: \mathcal{A} \to \mathcal{S}$ such that for each $B \in \mathcal{A}$, $H_A(B) = \mathrm{Hom}_{\mathcal{A}}(A,B)$. YL relates the sets $F(A)$ and $\mathrm{Nat}(H_A,F)$ for any $A \in \mathcal{A}$ and $F: \mathcal{A} \to \mathcal{S}$. The formulation of this relationship is given in [14] as follows:

<u>YL_1</u>. <u>For each</u> $A \in \mathcal{A}$ <u>and</u> $F: \mathcal{A} \to \mathcal{S}$ <u>there is a bijection</u>

$$\Psi_{A,F}: F(A) \cong \mathrm{Nat}(H_A, F) .$$

<u>This is explicitly defined for any</u> $x \in F(A)$, $B \in \mathcal{A}$ <u>and</u> $f \in H_A(B)$ <u>by</u> $(\Psi_{A,F}(x))(f) = (F(f)(x)$.

There is no special use of smallness here, besides that required to show $H_A: \mathcal{A} \to \mathcal{S}$. The canonical proof is readily formalized in ZFC($\underline{s}$).

The relationship in YL_1 is given uniformly by the explicit definition of $\Psi_{A,F}$. However, if one wants to express this uniformity by the notion of natural equivalence, it is necessary to consider the functor category $\mathcal{S}^{\mathcal{A}}$. The conventional formulation (e.g. in [1]) then requires $\mathcal{A}$ to be small. There is no need to assume this here, since functor categories can be formed without restriction as explained in §3.1.

<u>YL_2</u>. <u>Suppose</u> $\mathcal{F}$ <u>is any full category of functors from</u> $\mathcal{A}$ <u>to</u> $\mathcal{S}$, <u>containing</u> H_A <u>for each</u> $A \in \mathcal{A}$. <u>Let</u> $E: \mathcal{A} \times \mathcal{F} \to \mathcal{S}$ <u>and</u> $N: \mathcal{A} \times \mathcal{F} \to \mathcal{S}$ <u>be defined by</u> $E(A,F) = F(A)$ <u>and</u> $N(A,F) = \mathrm{Nat}(H_A,F)$ <u>for any</u> $A \in \mathcal{A}$, $F \in \mathcal{F}$. <u>Then there is a natural equivalence</u>

$$\Psi : E \cong N .$$

(It is understood that $E(f,\eta)$ and $N(f,\eta)$ are defined in the standard way.) Again this can be proved in ZFC($\underline{s}$). For any $\mathcal{A}$, $\mathcal{F} = \mathcal{S}^{\mathcal{A}}$ satisfies the hypothesis of YL_2. When $\mathcal{A}$ is small, the category $(\mathcal{S}^{\mathcal{A}})_{\underline{s}}$ of small functors from $\mathcal{A}$ to $\mathcal{S}$ also satisfies the hypothesis; it has already been shown to be full in § 3.1, and one easily checks that each H_A is small.

(iii) KE. The preceding shows that both AFT and YL (in both forms) are provable in ZFC($\underline{s}$) = ZFC + $\mathcal{R}\mathrm{In}_{\mathcal{L}}(\underline{s})$. Only KE, as currently stated, seems to require ZFC + RIn($\underline{s}$).[10] Let $\mathcal{A}$ be a small category, $\mathcal{B}$ a locally small category, $\mathcal{S} = \mathcal{S}_{\underline{s}}$, and let $J: \mathcal{A} \to \mathcal{B}$. This induces a functor $\mathcal{S}^J: \mathcal{S}^{\mathcal{B}} \to \mathcal{S}^{\mathcal{A}}$, with $\mathcal{S}^J(G) = JG$ for any G in $\mathcal{S}^{\mathcal{B}}$. With these hypotheses, KE asserts that

$$\mathcal{S}^J : \mathcal{S}^{\mathcal{B}} \to \mathcal{S}^{\mathcal{A}} \ \underline{\text{has both left and right adjoints}} .$$

The left adjoint associates a functor $K_F: \mathcal{B} \to \mathcal{S}$ (the Kan extension of F) to each $F: \mathcal{A} \to \mathcal{S}$ by the following construction for each $B \in \mathrm{Ob}_{\mathcal{B}}$:

$$K_F(B) = \varinjlim_{(J,B)} \overline{F} ,$$

where (J,B) is the category whose objects are all maps $J(A) \to B$ and whose morphisms are those corresponding to morphisms $A \to A'$,

---

10) I am indebted to Lawvere for bringing it to my attention for this reason. I have not studied the literature on KE, which I understand goes back to [9]. This has been explained to me in correspondence from Lawvere and discussion at the meeting.

and where $\overline{F}(J(A) \to B) = F(A)$. This direct limit can thus be obtained from $\coprod_{A \in Ob_{\mathcal{A}}} F(A)$. We cannot conclude that $K_F(B) \in \underline{s}$ unless we know that F <u>is small</u>. This is automatically satisfied if $In(\underline{s})$ holds. However, <u>if the only applications which are made of KE use K_F for $\mathcal{L}$-definable F, then the result has an adequate version in</u> ZFC + $RIn_{\mathcal{L}}(\underline{s})$. I do not have enough information about the uses of KE to judge this, but it is plausible on general grounds. Indeed, I would guess that the following weakening of KE is completely adequate:

<u>Suppose $\mathcal{A}$, $\mathcal{B}$ are small categories and</u> $J: \mathcal{A} \to \mathcal{B}$. <u>Then</u> $\mathcal{S}^J: (\mathcal{S}^{\mathcal{B}})_{\underline{s}} \to (\mathcal{S}^{\mathcal{A}})_{\underline{s}}$ <u>has both left and right adjoints</u>.

That is, one restricts attention throughout to <u>small functors</u>; this restriction is certainly coherent with the hypothesis.

3.4. <u>Case study of current category theory: uses of the reflection principle</u>. We now assume all of ZFC/$\underline{s}$. The following shows how the reflection principle and Theorem 1 can be applied to procedures in homological algebra [13], such as the calculation of the functors Ext^n. This is illustrated by the familiar case of Abelian groups. A,B,C,... range over arbitrary Abelian groups, and $(A \to B)$,... over Abelian group homomorphisms. One way of defining $Ext^n(C,A)$ is by an equivalence relation $\equiv$ between long exact sequences

$$E: 0 \to A \to B_1 \to \dots \to B_n \to C \to 0 .$$

An Abelian group structure is defined on any collection of representatives [E] of the equivalence classes; representatives must be chosen since the equivalence classes do not exist as sets. This would seem to require AC; however, the following general device, due to Scott, gives an explicit association serving the same purpose. Let [E] be <u>the set of all</u> E' <u>of least rank such that</u> E ≡ E'. With this definition the following can also be shown: (i) <u>for any</u> A, C <u>and</u> n <u>there is a group</u> $\mathrm{Ext}^n(C,A)$ <u>consisting of all</u> [E] <u>associated with exact sequences</u> E <u>of the above form</u>; (ii) <u>if</u> A, C <u>are small then each</u> [E] <u>is small and</u> $\mathrm{Ext}^n(C,A)$ <u>is small</u>.

The second method of defining Ext^n is via projective resolutions X, which are certain sequences

$$\ldots \to X_2 \to X_1 \to X_o \to C \to 0 .$$

Then $\mathrm{Ext}^n(C,A)$ is isomorphic to the nth cohomology group $H^n(X,A)$ for any such X. It can be shown that: (iii) <u>if</u> C <u>is small then there exists a small projective resolution</u> X <u>of</u> C, <u>and</u> (iv) <u>if</u> A <u>is small and</u> X <u>is small then</u> $H^n(X,A)$ <u>is small</u>.

Thus either way of defining Ext^n leads to a group construction whose restriction $\mathrm{Ext}^n_{\underline{s}}$ to small Abelian groups is a functor

$$\mathrm{Ext}^n_{\underline{s}}: \mathcal{AG}^{op}_{\underline{s}} \times \mathcal{AG}_{\underline{s}} \to \mathcal{AG}_{\underline{s}}$$

such that $\mathrm{Ext}^n_{\underline{s}}(C,A) = \mathrm{Ext}^n(C,A)$ for any small A, C. Now all of the work of homological algebra on Ext^n can be comfortably carried out for this functor on the category of small Abelian groups. One of the simplest results of this work is the statement

$$\forall C, A \in \underline{s}[\mathrm{Ext}^2_{\underline{s}}(C,A) = 0] .$$

It immediately follows by the reflection principle for <u>s</u> that

$$\forall C, A[\mathrm{Ext}^2(C,A) = 0] .$$

Then, by Theorem 1, the latter is also provable in ZFC.

The treatment just indicated is a special case of the following general procedure. Let M(x), C(x,y,z) be formulas of $\mathcal{L}$ for which we can prove that the conditions to be a category are satisfied when we take the morphisms to be those x such that M(x), and the composition relation $xy = z \leftrightarrow C(x,y,z)$.[11] In this case we speak of the "<u>meta</u>-<u>category</u>" $\mathcal{A}$ determined by the <u>properties</u> M,C. $\mathcal{A}$ cannot in general be identified with a set. However, the category $\mathcal{A}_{\underline{s}} = (M_{\underline{s}}, C_{\underline{s}})$ with $M_{\underline{s}} = \{x | x \in \underline{s} \wedge M(x)\}$, $C_{\underline{s}}(x,y) = z \leftrightarrow x,y,z \in \underline{s} \wedge C(x,y,z)$, <u>is</u> a set. The reflection principle for <u>s</u> tells us, roughly speaking, that $\mathcal{A}_{\underline{s}}$ <u>and</u> $\mathcal{A}$ <u>have exactly the same properties expressible in the language</u> $\mathcal{L}$ <u>of set theory</u>. Assuming § 3.2 II, all of current categorical and homological algebra can be applied to any such category $\mathcal{A}_{\underline{s}}$. It then follows by Theorem 1 that <u>each specific result about</u> $\mathcal{A}_{\underline{s}}$ <u>formulated in</u> $\mathcal{L}$ <u>automatically yields the same result for</u> $\mathcal{A}$ <u>in</u> ZFC.

Whether or not one is surprised by the possibility of such a general procedure, one should not be surprised by the consequences. These simply bear out the feeling that the notion of smallness is a device for legitimizing some general constructions, without in

---

11) Examples besides those already given are provided by formulas M(x) expressing that x is a homomorphism between R-modules, x is a continuous map between topological spaces, etc.

any way affecting the consequences formulated in ordinary mathematical terms.

4. Open problems.

Naturally further work is needed to verify the impression (§ 3.2II) that current category theory can be formulated in ZFC + $\mathrm{RIn}_{\mathcal{L}}(\underline{s})$ (it is to be expected that strong use will be made of the reflection principle). In particular, it should be seen whether the weakened version of KE given in § 3.3(iii) is a completely adequate version of KE.

4.1. Model-theoretic problems. Model theory treats properties of structures, e.g. of the form $\mathfrak{N} = (N,R)$ where N is a set and $R \subseteq N^3$, given by formulas in specified languages. Usual 1st order model theory takes the language $\mathcal{L}_{\omega,\omega}$ with variables ranging over N, basic symbols for = and R, and finitary connectives and quantifiers. More recently, stronger languages have been considered, e.g. $\mathcal{L}_{\omega_1,\omega}$ with countably long conjunctions and disjunctions, and $\mathcal{L}_{\omega_1,\omega_1}$ with countably long quantifier sequences in addition.

$\mathfrak{N}$ and $\mathfrak{N}'$ are said to be equivalent in a given language if they satisfy the same sentences in that language. Much work in model theory deals with problems concerning these equivalence relations, in particular for various classes of algebraic structures (cf. e.g. [19]). The same sorts of problems can be considered for categories $\mathcal{a}$ taken as structures $(M_{\mathcal{a}}, C_{\mathcal{a}})$. For example, for which cardinals γ is the category $\mathcal{a}\mathcal{B}_{\underline{s}}$ of all small Abelian groups

<u>equivalent in</u> $\mathcal{L}_{\omega,\omega}$ ($\mathcal{L}_{\omega_1,\omega}$, etc.) <u>to the category</u> $\mathcal{A}\mathcal{B}_\gamma$ <u>of Abelian groups of cardinality</u> $<\gamma$? This may tie up with 5.1; in some cases one should be able to find closure conditions on a category which completely determine all its properties in a given language (i.e., give a complete theory).

Note that many properties of a category dealt with in homological algebra, say those involving arbitrary finite exact sequences, can be formulated in $\mathcal{L}_{\omega_1,\omega}$; further properties, e.g. those concerning infinitely long exact sequences can be formulated in $\mathcal{L}_{\omega_1,\omega_1}$. Thus the model-theoretic questions raised should be connected with problems of mathematical interest. Note also that in ZFC/$\underline{s}$ the equivalence with respect to sentences of $\mathcal{L}$ of a metacategory $\mathcal{A}$ and $\mathcal{A}_{\underline{s}}$ (§3.4), implies in a suitable sense the equivalence of $\mathcal{A}$ and $\mathcal{A}_{\underline{s}}$ in $\mathcal{L}_{\omega,\omega}$. It can also be shown to imply the equivalence of $\mathcal{A}$ and $\mathcal{A}_{\underline{s}}$ for a wide class of sentences in $\mathcal{L}_{\omega_1,\omega_1}$.

4.2. <u>Axiomatization of category theory</u>. To make a presentation of category theory intelligible it is evidently not sufficient (and hardly necessary) to define its basic notions set-theoretically. The subject becomes clear, like any other branch of mathematics such as number theory, only if it is formulated in a language proper to it. In the case of number theory, one uses variables ranging over the natural numbers, in category theory over categories and functors. One language for category theory has been proposed by Lawvere [10] in his elementary theory ETCC of the "category of all categories"; we shall call the language CT (and reserve 'ETCC' for the axiomatic system). Let ZFC/s-CT consist of the statements of this language whose set-theoretical translation is provable in ZFC/s. There are two problems: (i) What is a simple set of axioms for ZFC/s-CT? (ii) What is the relationship of ETCC to ZFC/s-CT? Of course, similar questions can be raised for the theories extracted from other systems of set theory. If I understand ETCC properly, it seems that it can be interpreted in ZC + RIn(s), where Z = Zermelo's axioms; however, I have not worked out the details. Note that ZC + RIn(s) proves the existence of a model for ZFC and so is stronger than ZFC.

5. Further work.

5.1. Analyzing closure conditions. Each particular result in ZFC + $RIn_{\mathcal{L}}$(s) make use of only a finite number of statements from $RIn_{\mathcal{L}}$(s), expressing that s is closed under certain operations. Thus each result involving the collection of small sets generalizes to a result of the form: for any collection of sets satisfying such and such closure conditions, we have ... . In category theory this will lead to statements of the form: for any category of sets (category of functors, etc.) satisfying such and such closure conditions, we have ... . These statements no longer involve any unusual mathematical notions, and can be formulated and proved by ordinary mathematical means without special reference to any formal system. The following conjectures were made by Kreisel in the notes mentioned in ftn. 3: For the bulk of theorems

<u>in category theory (in contrast to, say, current set theory) very weak closure conditions on the families of sets and functors involved are needed; furthermore it is mathematically fruitful to state explicitly such closure conditions</u>. Mac Lane's axioms for schools [16] can be regarded as providing closure conditions sufficient for the AFT; however, they are not sufficient for other theorems. I have carried out work illustrating Kreisel's conjectures for YL as well as AFT[11], making use of the types of relativizations indicated in § 3.1.

5.2. <u>Definable categories and functors</u>. The results of category theory are eventually applied to members of various metacategories, e.g. of all R-modules, all topological spaces, etc. The general arguments and constructions are canonical in nature and always quite explicit. Thus it would seem that nothing would be lost if attention were restricted to (§) <u>definable</u> categories, functors, etc. throughout. Or, more conservatively, it should be possible to capitalize on such a restriction. Kreisel has sketched, in the Seminar notes mentioned in ftn. 3), one way of doing this in the BG theory of classes (§ 6). This formulation provides direct means to treat cases of self-application, such as a semi-group structure on all semi-groups, a category structure on all categories, etc. It should be quite interesting to pursue this and to investigate its relationships with the formulation studied in this paper.

---

11) In unpublished notes for the Seminar mentioned in ftn. 3); to appear in the sequel to this paper.

APPENDIX I

6. Formulation of category theory in two theories of classes.

The language $\mathcal{LC}$ has two sorts of variables, those ranging over sets a,b,c,...,x,y,z,... and those ranging over classes A,B,C,...,X,Y,Z,...; it is otherwise the same as the language $\mathcal{L}$ of §2.1.

The following system of axioms BG is equivalent to the system of Bernays-Gödel developed in [3].

(0) $\forall x \exists X(x = X) \wedge \forall X[\exists Y(X \in Y) \rightarrow \exists x(x = X)]$.

(i) - (vii) of ZF

(viii) $\forall a \forall X\{\forall x,y_1,y_2[\langle x,y_1\rangle \in X \wedge \langle x,y_2\rangle \in X \rightarrow y_1 = y_2] \rightarrow$

$$\exists b \forall y[y \in b \leftrightarrow \exists x(x \in a \wedge \langle x,y\rangle \in X)]\}$$

(ix) for each formula $\varphi(x)$ with free variables $x,y_1,\dots,y_n,Y_1,\dots,Y_m$ altogether, but with no class quantifiers,

$\forall y_1,\dots,y_n \forall Y_1,\dots,Y_m \exists X \forall x[x \in X \leftrightarrow \varphi(x)]$.

As presented by Gödel [3], there is only one sort of variable ranging over classes. A class X is defined to be a set if $\exists Y(X \in Y)$; then set variables are introduced by convention. In this case axiom (0) is dispensable. Furthermore, one has only finitely many instances of (ix) in Gödel's system. It is shown in [3] how to derive every instance of (ix) from this finite system of axioms.

It is obvious that every axiom of ZF is derivable in BG. Hence if (M,C,E) is a model of BG, where M is the collection of 'sets' and C the collection of 'classes' then $(M, E \upharpoonright M)$ is a model of ZF. On the other hand, if (M, E_1) is a model of ZF then (M, D_M, E) is a model of BG, where D_M consists of all subsets X of M definable in the form

$$X = \{x \mid \varphi(x, a_1, \ldots, a_n) \text{ holds in } (M, E_1)\}$$

for φ any formula of $\mathcal{L}$ and $a_1, \ldots, a_n \in M$ (each <u>a</u> in M is identified with the set $X = \{x \mid xE_1a\}$); the relation E is the extension of E_1 to D_M by the membership relation $\in$. It follows that BG is a <u>conservative</u> extension of ZF.[13)]

This permits one to obtain metamathematical results about ZF, for example concerning consistency and independence of sentences of $\mathcal{L}$, from corresponding results for BG. The finite axiomatizability of BG can have some technical advantages, as in [3]. In addition, BG is of foundational interest because it can be thought of as <u>a theory of sets and (extensions of) set-theoretically definable properties</u>.

The weakness of the class-existence axioms (ix) is familiar. What is less well known is that there are instances of the schema of ordinary mathematical induction,

$$\text{(MI)} \qquad \varphi(0) \wedge \forall x[\varphi(x) \rightarrow \varphi(x')] \rightarrow \forall x \in \omega\, \varphi(x)$$

(φ an arbitrary formula of $\mathcal{LC}$) which are not derivable in BG. This was shown by Wang [20], roughly as follows. There is a

---

[13)] This was first pointed out by Mostowski [18], where the elementary nature of the above argument, and hence of the relative consistency of BG to ZF was stressed (it can be formalized in a weak subsystem of ZF).

formula $\Psi(x)$ of $\mathcal{LC}$ which expresses for $x \in \omega$ that x is the Gödel number of a true sentence of $\mathcal{L}$. One can prove in BG + MI that if x is the number of a sentence derivable from ZF then x is true. Hence the consistency of ZF is provable in BG + MI, and from this also the consistency BG is provable (cf. ftn. 13)). Then MI is not derivable from BG by Gödel's 2nd incompleteness theorem (given that BG is consistent).

A stronger system MK of sets and classes, due to Morse, is considered in Kelley [4]. It has the axioms (0)-(viii) of BG but now (ix) for <u>any</u> formula φ of $\mathcal{LC}$. Since

$$BG \vdash \forall X\{0 \in X \wedge \forall x[x \in X \rightarrow x' \in X] \rightarrow \forall x \in \omega (x \in X)\} ,$$

we have all instances of MI provable in MK.

In contrast to BG, MK is not finitely axiomatizable, and we cannot interpret the classes as ranging over a collection of properties defined in advance. MK has a model in any $(R_\alpha, R_{\alpha+1}, \in\upharpoonright R_{\alpha+1})$ for α strongly inaccessible. It follows that MK can be interpreted in ZF + RIn(<u>s</u>).

Eilenberg and Mac Lane proposed a formulation of category theory in the language $\mathcal{LC}$, translating 'small' and 'large' by 'set' and 'class', respectively. This was developed in Mac Lane [12] and has since been applied extensively in practice, though not without reservations. The mathematical defects of this proposal are familiar; for example, one cannot deal with functor categories $\mathcal{B}^{\mathcal{A}}$ except when $\mathcal{A}$ is small.

No explicit hypothesis was advanced in [12] about which formal system in $\mathcal{L}\mathcal{C}$ would be adequate to current category theory. However, one has the impression that BG together with AC was considered for this purpose, since this is the only system which has been mentioned in this connection. In view of the inadequacy of BG to the full schema MI, there is reason to be cautious about such an hypothesis. On the other hand, the evidence for the following is as strong as for Hypothesis I of § 3.2:

Hypothesis III. Every theorem of current category theory is provable in MKC.

By the above, the Hypothesis I is established if III is accepted. The question of interest here is whether 'MKC' can be replaced by 'BGC' in III. This would be established by Hypothesis II and Theorem 1, but perhaps also more directly by other means as indicated in § 5.2.

APPENDIX II

by G. Kreisel

7. Category theory and the foundations of mathematics.

Though the set-theoretical foundations described above use only standard methods of mathematical logic, they involve some points which are not familiar to the *general* mathematical public. It may therefore be useful to say a word on these matters and to give references to the literature where more information can be found.

7.1. Basic set-theoretical notions. Historically, there are two outstanding facts. First, we have the very well-known Russell paradox which shows that the first attempt, by Frege, of formulating set-theoretical principles was defective; more precisely, as we should say now (see [7] pp. 171-173 for the distinctions needed), Frege attempted to formulate axioms for a very abstract notion of 'property', not of 'set' as used in mathematics. Second, we have the much *less dramatic* and therefore less well-known fact that Frege's contradictory axioms are not even remotely plausible if one only stops to ask: what is a set? (though it is perhaps not 'natural' to ask the question). And the axioms are evidently false if one gives the specific answer, of Zermelo [21], that the sets to be considered are built up by iterating the *power set operation* P:

$$a \mapsto \text{collection of all subsets of } \underline{a} .$$

(Actually Zermelo found a smoother theory by considering instead the

<u>cumulative</u> operation R: a $\mapsto$ a $\cup$ Pa.) In a sense to be explained in a moment, it is 'sufficient' for mathematics to consider sets built up from the <u>empty set</u>, that is

$$R_0 = 0 \text{ and, for } 0 < \beta \leq \alpha,\ R_\beta = \bigcup_{\gamma < \beta} R(R_\gamma),$$

where α is an <u>ordinal</u>.

Paradoxically, though most mathematicians have heard criticisms of the power set operation, they are unaware of the <u>genuinely</u> problematic matter of <u>how often the operation</u> R <u>is to be iterated</u>. (There is evidently no biggest α; for more detail, see [6].) One reason for this ignorance is accidental; when the paradoxes were first discovered, people were 'thrashing about' for solutions. Perhaps the first feature of P to come to mind was that, for infinite <u>a</u>, the elements of Pa could not all be listed (or 'defined', though people rarely stopped to think what means of definition were to be used, see [6]I). This was considered to make the use of the power set operation 'illegitimate' and so, naturally, one never got off the ground; one certainly did not consider the iteration of P. A more objective reason, and one of possible interest to category theorists, is this:

> The classical mathematical structures (arithmetic, continuum) occur, that is have isomorphic copies, in quite 'short' segments of the cumulative hierarchy, certainly in R_α, for $\alpha < \omega + \omega$, and so the question of exorbitant iterations simply did not arise.

Whatever doubts one may have about the iteration process, in fact about the meaning of 'ordinal', they do not affect these short segments of the hierarchy. The definition can be rewritten, using only numerical variables n (not for infinite ordinals)

$$R_o = 0,\ R_{n+1} = R_n \cup P(R_n);\ S_o = \bigcup_n R_n,\ S_{n+1} = S_n \cup P(S_n) .$$

In terms of familiar axiomatic systems in the language of set theory, as in the text above, we have this: When the classical structures are defined set-theoretically, their basic properties are generally established using only the principles of Zermelo's original (1908) set theory. These axioms are evidently satisfied if the variables are taken to range over $\bigcup_n S_n$ (or, generally, R_α for limit ordinals α).[14)]

In contrast, the reflection principle, used in an essential way in the set theoretical foundations of category theory given above, is evidently not satisfied in $\bigcup_n S_n$, nor, roughly speaking, in any 'short' segment. (One can be more precise here by considering second order versions of set theory; see Montague and Vaught [17] concerning the first order and second order replacement axiom. The latter is satisfied in R_α only for inaccessible α, the former for suitable countable α, but the natural proof of this last fact uses the existence of an inaccessible α.)

For reference in 7.3 below: the reflection principle is not a formal consequence of the replacement schema (and the remaining axioms of ZFC) without the axiom of foundation. This formal

---

14) R_α corresponds to the universe U in the introduction.

fact is of no importance for the sets we are talking about because the cumulative hierarchy _is_ well-founded. But it stands out in actual practice because the axiom of foundation is rarely involved in ordinary mathematics.

Digression on the system BG, described in §6. This formal system did not, so to speak, drop from heaven (into an empty space of von Neumann's head, whose formulation was modified by Bernays and finally by Gödel to yield BG). It was obtained in order to give a finite axiomatization of ZFC, specifically to replace the _schemata_ of ZFC by a finite number of conditions. The schemata, for instance the comprehension schema, arise by weakening a second order axiom of the form: for _every_ property (of elements of a set _a_) there is a set such that ..., to: for every property _definable_ in the language of ZFC there is a set such that .... Put in terms of the cumulative hierarchy, if R_α is a model of ZFC (with the universe R_α and '$\in$' interpreted as membership restricted to R_α) then the collection $R_{D,\alpha+1} \subset R_{\alpha+1}$ is a model of BG, where $R_{D,\alpha+1}$ consists of those sets which are explicitly definable by formulas of ZFC over the universe R_α; constants for elements $\in R_\alpha$ are allowed in these definitions.[15)]

7.2. _Adequacy conditions for set-theoretical foundations_ (see [7] pp. 167-168).

In 7.1 and in the main text of this paper, vague terms like 'sufficient' or 'adequate' (foundations of mathematical practice)

---

15) Note that $R_{\alpha+1}$ itself is not a model of BG unless α is inaccessible.

were used. For set-theoretical foundations as _originally_ intended, it is easy to formulate precise adequacy conditions: we suppose given an intuitively understood notion, either of a specific structure like that of arithmetic or of a property like transitive relation or functional relation, and then we try to characterize the notion up to isomorphism in set-theoretical terms. For instance, the remark in 7.1 above that it is 'sufficient' to consider the hierarchy R_α means this. For the intuitive structures considered, we can explicitly define an isomorphic copy in some R_α; specifically we do not need to consider (as Zermelo originally did) a hierarchy $R_\alpha[a]$ and a language with new relation symbols $p_1,\ldots,p_n$, say, in addition to $\in$, where $R_0[a] = a$, _a_ a collection of individuals, that is objects without members, $p_i (1 \leq i \leq n)$ denoting relations $\bar{p}_i$ defined on _a_. We do not 'need' means that the structures of mathematical practice happen to be definable in set theory itself. Naturally, we may need a _longer_ segment R_α before getting an isomorphic copy of a given structure, e.g., trivially, for an infinite _a_, the structure $(a,\bar{p}_1,\ldots,\bar{p}_n)$ itself will occur only in R_α for a suitable _infinite_ α, while it occurs in $R_m[a]$ for a suitable _finite_ m (with the usual coding of n+1-tuples).

The pioneers of mathematical logic provided set-theoretical characterizations of basic mathematical notions. Frege and Dedekind gave their famous axioms for specific structures, the natural numbers (with first element and successor) and the continuum respectively; Whitehead and Russell for many properties of what we should now call relational systems, e.g., total or partial ordering.

Going back to the intended interpretations of ZFC, namely segments R_α, the specific classical structures are elements of R_α or, in terms of [7], their axioms have a set-theoretical realization in the strict sense since the universe $\in R_\alpha$; in contrast, the (first order) properties of relational systems mentioned above correspond to a generalized realization because no element of R_α is the set of all relational systems ($\in R_\alpha$) which are orderings.[16)]

More recently, people have studied systematically different kinds of characterizations among set-theoretical ones, corresponding to the order of the language needed for the characterization of a particular structure or notion. Famous results here show that no infinite (specific) structure can be characterized by first order axioms, nor, e.g., the (non-categorical) property of being a well-ordering. These results establish the inadequacy of 'first order' foundations. This series of researches is a good example of foundational research; in particular it gives an idea of the kind of problems that arise in set-theoretical foundations tout court, that is without the restriction to 'first order'.

Let us note in passing that the classical adequacy conditions discussed in the present section are not sufficient to decide between a foundation by the hierarchy R_α of 7.1 and (Zermelo's original form) $R_\alpha[a]$ in 7.2. (A formal difference between them is that R_α satisfies extensionality and $R_\alpha[a]$ does not for $a \neq 0$.) Indeed,

---

16) Also the second order property of a well ordering; as Burali-Forti knew the generalized realization of all well-orderings in R_α with the relation $\leq$ is again a well-ordering, but it is not an element of R_α. ($\lesssim$ means: there is an order preserving injection.)

when we think of actual practice with its unlimited 'supply' of individuals a variant $R'_{\alpha}[a]$ seems even more appropriate: Instead of the power set operation P, we consider P':

$x \mapsto$ collection of all subsets of x hereditarily of cardinality $<$ card(a);

define $R': x \mapsto x \cup P'x$, and build up $R'_{\alpha}[a]$ analogously to $R_{\alpha}[a]$ with R' replacing R. It remains to be seen whether this impression (of appropriateness) can be justified by a more detailed analysis of adequacy conditions.

7.3. Set-theoretical foundations of category theory. Though objectively not very important, it will help avoid misunderstandings if we reflect why logicians found it difficult to take seriously the problems of foundations for category theory as stated in the literature. On the one hand, we found a perfectly straightforward property of relational systems, as in 7.2 above, of 'being a category'; it is a first order property and so, since it is not empty, it only has a generalized realization; it happens to permit self-application as the property of being a well-ordering in footnote 16. On the other hand, we were told that mathematicians 'want' or 'need' to use illegitimate totalities, and that it was somehow a defect of set theory that stood in the way.[17] There certainly was no obvious difference between the proposed uses and that of Burali-Forti in footnote 16 except that he, as Frege before him, at least

[17] During the war I knew a student at Cambridge (not a great mathematician) who handed in an exercise consisting of two parts: (i) Prove that $-1 \leq \cos \theta \leq 1$; he did this; he began his proof of (ii) by 'let $\cos \theta$ tend to infinity! He needed this property for his proof of (ii).

came right out and stated a principle he wished to assert about the notions considered. Zermelo, in 1908, may have been too scathing about the thoughtlessness of the people who got themselves involved in the paradoxes. But 50 to 60 years later we, that is the logicians, were naturally suspicious of proposals which simply disregarded the literature on the very question at issue; not side remarks by obscure authors but pellucid and forceful non-technical discussions by Zermelo himself or Gödel (for references, see [6]). These suspicions were equally natural whether one assumed that category theorists simply ignored this literature or that they had not understood it (see also 7.6 below, particularly footnote 19).

What was missing before one could even begin to consider a problem of foundations for category theory, was a formulation of the latter. In other words, the specialists in the subject should decide what they are talking about, and state properties of these concepts (it is the business of specialists because 'category' in contrast to 'number' was introduced as a technical notion). Then we can see whether or not we get an adequate set-theoretical foundation. Problem (i) of 4.2 is so to speak the reverse: from a possibly quite vague impression of the general practice of category theory, one picks on a language CT proper to the subject, and then defines the primitive terms of CT set-theoretically in ZFC/s where the notion of 'small' is a new primitive.[18)]

[18)] I do not feel at all competent to judge the value of such work for category theory itself. But there is no doubt in my mind of its value in connection with a more abstract problem. Though properties, in the sense of 7.2, only have generalized realizations,

The adequacy established for this reduction of CT to set theory via ZFC/s is not given by an isomorphism result, but only by a conservative extension result.

Summarizing: there is no evidence that the self-reflexive operations of category theory introduce anything radically novel over those known since the beginning of set-theory (well-ordering of well-orderings, semigroup structure of semigroups etc.); but if the use of the reflection principle in the present paper should turn out to be essential to the development of category theory then, by 7.1 above, the set-theoretical foundations of category theory really involve properties of sets not used in most other branches of mathematics.

The remaining passages concern non-set-theoretical foundations and therefore are not directly relevant to the present paper. But recent literature on the foundations of category theory flirts with some non-set-theoretical traditions, and it is more satisfying to go the whole hog.

7.4. Formalist foundations. A simple-minded, but rather common misunderstanding is the belief that the explicit formulation of notions and principles is a prerogative of formalist foundations.

---

can we not find particular concrete representatives at least within set-theoretical contexts? The answer is: for set-theoretically defined properties φ (defined for the whole universe), the object $\{x \in s: \varphi^{(s)}(x)\}$ is such a representative. (Compare here a von Neumann ordinal α which is a representative for the property of being a well ordering of ordinal α, with respect to all properties invariant under isomorphism.) From the present point of view the letter 'r' would be better than 's'.

The essential point of formalism is that _certain_ formulations are rejected, for instance second order formulations involved in such notions as 'compactness' as ordinarily understood (specifically, where the notion of _subset of a space_ is understood). The only notions to be used refer to combinatorial operations, in particular those used in defining and manipulating formal systems. Abstract notions are not included, only formal operations on words or formulae used to represent assertions about such notions.

Evidently for formalist foundations totally different adequacy conditions are required from those in 7.1. As is well known proper conditions including that ofacombinatorial consistency proof were formulated by Hilbert; for a pedantic exposition see e.g. [7] App II B. A particular source of confusion is that the reduction of mathematical reasoning to set theory plays a role in formalist foundations too. However, in contrast to 7.1, not characterizations up to isomorphism are involved, but the _formal_ reduction of _formalized_ mathematical reasoning to _formalized_ set theory; the latter serves as a compact description of mathematical practice.

In contrast to the preoccupation of the present paper with eliminating the assumption of inaccessibles, for formalist foundations in the sense of Hilbert, _a much more radical requirement is needed_: a reduction to set theory is useful, that is ensures a consistency proof by the means desired, only if, roughly speaking, we also eliminate the power set axiom and even applications of the comprehension schema to formulas above a quite low level of _logical complexity_.

As already stressed in 7.1 such restrictions <u>on</u> the axioms would be quite out of place in set-theoretical foundations, where, on the contrary, the restriction <u>to</u> properties definable in the language of ZFC (by formulae of arbitrary complexity) is <u>ad hoc</u>. Evidently the different role of formally similar principles in two different foundational schemes is a source of confusion.

7.5. <u>New foundations</u>. There is a well-known platitude that future developments of category theory (or number theory using large cardinals) may require revision of current foundations. Here it is to be remembered that we know anyway several natural, perfectly traditional notions with a mathematical 'flavour' which present serious problems for set-theoretical foundations <u>if we seriously wanted to reduce these notions to set-theory</u> (certainly, they haven't been reduced). One example is the general notion of <u>rule</u> treated in intuitionistic foundations, where a rule is regarded as a <u>process</u>; in particular it is not identified with its graph because in fact a rule may be understood before or without any clear knowledge of its domain, for instance the identity operation. Another example is that of <u>abstract structure</u> or <u>abstract property</u> which, despite footnote 18, is only partially expressed in set-theoretic terms. If one seriously considered non-set-theoretical foundations, one would begin with the (primitive) notions of <u>rule</u> and <u>abstract structure</u> in terms of which the notions of category theory can be defined; and not, of course, with the technical notion of category itself.

But granted all this, practically speaking the following point is much more important. What we know or can say about these notions is simply not comparable in clarity, generality or elegance to what we know of the segments of the cumulative hierarchy which are relevant to set-theoretical foundations of mathematical practice. I believe one of the worst effects of uninformed criticism of set-theoretical foundations, as of the kind mentioned in 7.1 or 7.3, is that it has blinded people to the standards of rigour that have already been attained in foundations.

7.6. Foundations and mathematical practice. The reader will have heard such expressions as 'foundations of ring theory', not in the sense of a logical analysis in terms of some foundational scheme, but simply as an organization or presentation of the subject. The expression is not necessarily intended as a pun but corresponds to a positivistic conviction which became current after the failure of Hilbert's programme. Forgetting that the latter was intended to establish a really quite implausible conjecture (namely the possibility of a formalist reduction of mathematical reasoning) people thought there was no hope of any foundational analysis! They thought that the nearest one could get to in this kind of direction was: organization.[19] Odd as this reaction may be (for instance,

[19] On this same view the discovery of axioms is supposed to be made by describing what mathematicians 'do' and not by analyzing concepts. In particular, the basic adequacy condition, of 7.2, on the characterization of intuitively understood notions (as in the characterization of arithmetic by means of Peano axioms) would be rejected. The view is most unempirical if one remembers how axioms were actually found! Naturally, if one accepts the view, one does not accept the difference, stressed in the second paragraph of 7.3, between axiomatizing 'number' and 'category': this ultimately leads to the state of affairs criticized in paragraph 1 of 7.3.

Gödel's incompleteness theorem does not constitute the least difficulty for set-theoretical foundations in the sense of 7.1), there it is and we had better be aware of it.

Organization and foundations are incomparable. Organization involves a proper choice of language; we have already seen (4.2 or 7.3) that this is not necessarily provided by set-theoretical foundations. On the other hand, we may have a very successful organization which leaves open the verification of adequacy conditions, at least for a given foundational scheme.

Generally speaking, the aims of foundations and organization will be in conflict. Being an analysis of practice foundations must be expected to involve concepts that do not occur in practice (just as fundamental theories in physics deal with objects that do not occur in ordinary life). Organization is directly concerned with practice; compare here the remark in 7.1 concerning the axiom of foundation.

But occasionally foundations can have heuristic value for practice. For example, (at least I hope it is a good example) the project of 'algebraization' in 6.1, that is of analyzing closure conditions systematically, came from experience with formalist foundations (see end of 7.4) which shows that surprisingly simple instances of the comprehension schema are needed for formalizing the bulk of actual practice. And, more specifically, the proposal of considering indexed functor categories (mentioned at the end of 3.1) came from familiar models of the 'subsystems' of set theory used in such formalizations. The best known of these models are short segments of the ramified or, as it is now known, constructible hierarchy.

BIBLIOGRAPHY

[1] P. Freyd, Abelian categories, Harper and Row, New York, 1964.

[2] P. Gabriel, Des catégories abéliennes, Bull. Soc. Math. France., v. 90 (1962) pp. 323-448.

[3] K. Gödel, The consistency of the axiom of choice and the generalized continuum hypothesis, Annals of Math. Studies, V. 3, Princeton, 1951.

[4] J. L. Kelley, General topology, Van Nostrand, Princeton.

[5] G. Kreisel, Mathematical logic, pp. 95-195 in Lectures on Modern Mathematics, vol. III (ed. T.L. Saaty), Wiley, New York, 1965.

[6] ________, Two notes on foundations, to appear in Dialectica.

[7] G. Kreisel and J.-L. Krivine, Elements of mathematical logic; model theory, North-Holland, Amsterdam (1967).

[8] J.-L. Krivine, Theorie axiomatique des ensembles, Presses Univ. de France, Paris, 1969.

[9] W. F. Lawvere, Functorial semantics of algebraic theories, Dissertation, Columbia University, 1963.

[10] ________, The category of categories as a foundation for mathematics, pp. 1-20 in Proc. Conf. on Categorical Algebra, La Jolla 1965, New York, 1966.

[11] A. Lévy, Axiom schemata of strong infinity in axiomatic set theory, Pacific J. Math. v. 10 (1960), pp. 223-238.

[12] S. Mac Lane, Locally small categories and the foundations of set theory, pp. 25-43 in Infinitistic methods, Oxford, 1961.

[13] _______, Homology, Springer, Berlin, 1963.

[14] _______, Categorical algebra, B.A.M.S. v. 71 (1965), pp. 40-106.

[15] _______, Foundations of mathematics: category theory, pp. 286-294 in Contemporary Philosophy, A Survey, I, Logic and Foundations of Mathematics, R. Kiblansky, editor, Florence 1968.

[16] _______, Foundations for categories and sets, Proc. 1968 Conference on Category Theory and its applications, Battelle Institute, to appear.

[17] R. Montague and R. L. Vaught, Natural models of set theories, Fund. Math. v. 47 (1959), pp. 219-242.

[18] A. Mostowski, Some impredicative definitions in the axiomatic set-theory, Fund. Math. v. 37 (1950), pp. 111-124.

[19] A. Robinson, Introductio
mathematics of algebra,

[20] H. Wang, Truth definitio
A.M.S. v. 73 (1962), pp.

[21] E. Zermelo, Über Grenzzahlen und Mengenbereiche, Fund. Math. v. 16 (1930), pp. 29-47.